Francis D. Conley

DATE DUE

EMCO

ALSO BY GERARD PIEL:

Science in the Cause of Man

THIS IS A BORZOI BOOK
PUBLISHED IN NEW YORK
BY ALFRED A. KNOPF

The Acceleration
of History

The Acceleration
of History

GERARD PIEL

Alfred A. Knopf
New York   1972

ISBN: 0-394-47312-4
Library of Congress Catalog Card Number: 71-171118

Several of the pieces in this volume appeared originally as follows: "The Acceleration of History" and "Peace and Quiet on the Campus" as pamphlets by Scientific American; "Last Words from the Dinosaurs" in The Explorer, November–December 1963, The Cleveland Museum of Natural History; "The Heritage of Science in a Civilization of Machines" in the Proceedings of the Sesquicentennial Convocation of Colby College; "On Recollection" in a pamphlet by the Smith College Library; "King Solomon's Ant" (originally entitled "The Comparative Psychology of T. C. Schneirla") in DEVELOPMENT AND EVOLUTION OF BEHAVIOR: ESSAYS IN MEMORY OF T. C. SCHNEIRLA, edited by Lester R. Aronson, Ethel Tobach, Daniel S. Lehrman, and Jay S. Rosenblatt, W. H. Freeman and Company, Copyright © 1970; "Why No Satisfaction" (originally entitled "The Advent of Abundance") in Science and Public Affairs, the Bulletin of the Atomic Scientists, Copyright © 1963 by the Educational Foundation for Nuclear Science; "Science Can Never Be Retrograde" and "Metropolitan Medicine" (originally entitled "Physician Heal Thy Society" and "Coming Changes in the Delivery of Medical Care in a Large Urban Health Center") in the Bulletin of the New York Academy of Medicine; "The Sorcerer's Apprentice" in the Proceedings of the 1964 Spring Joint Computer Conference of the American Federation of Information Processing Societies; "Getting Our Knowledge Fully Applied" in Proceedings of Subcommittee on Government Research, U.S. Senate; "The Grain That Steel Might Grow" in Nature; "The Treason of the Clerks" in the Proceedings of the American Philosophical Society; "Federal Funds and Science Education" in the Proceedings of the Committee on Science and Astronautics, U.S. House of Representatives; "The Subterfuge of Armament" (originally entitled "The Fork in the Road") in the Harvard Alumni Bulletin, July 7, 1962, Harvard Bulletin Inc.; "The Stationary State" in the Radcliffe Quarterly.

Manufactured in the United States of America
FIRST EDITION

for Eleanor Jackson Piel, Jr.

Contents

The Acceleration
of History

Afterthought

THE acceleration of history has brought the human species to the fork in the road. One road from here leads to a dead end. On the other, less plainly marked, our species may yet find the way to realization of its humanity.

If history is indeed to end in a suicidal conflagration, it will be from habit. Civilized communities have always employed force and violence, tempered at times by fraud, to hold their members together against the inequities that divided them. Inequity has had its historic role as well. It freed a few members of the community from brute toil throughout the ages when there never was enough to go around. The enormous forces and the violence now at the disposal of national states make this way of organizing society and going on with history suddenly fatal.

At the same time and by exercise of the same supremely human capacities that made the weapons of mass destruction, our species has brought another future into sight. Understanding and control of nature—science and technology—have shown how to secure enough-and-to-spare of material goods to permit the admission of all men into full membership in mankind. Inequity is thereby denied its historic sanction. Force and fraud lose, correspondingly, their acceptability as the tools of domestic management and international politics. The free and just society and a world at peace become practicable goals for individual and collective striving.

Over the past twenty-five years, the United States and its people have held out to mankind both the threat and the promise of the future in their popularly most intelligible models: Americans are well known to be in possession of the most abundant economy and the most destructive military machine history has ever known.

This country's abundance has signaled to the rest of the world the end of the regime of scarcity. At home it has reduced poverty among the people to an irreducible minimum; that is: not to be reduced by any further increase in abundance. Poverty in the United States is now a social rather than a material estate. It is inequity imposed on a minority of the people by the country's institutions, laws, and customs.

Such working of old compulsions in collision with new realities gives rise to pathology in other lines of behavior. Political heresy has legal standing in America. The economy is permitted to falter and throttle down at the cost of growth and promise to millions of the country's children and of privation and humiliation to millions of elders. In the wasting and curtailing of public amenities and services the private satisfactions of even the best-off are spoiled.

The same pathology underlies this country's commitment of its power abroad to history's dead end. No objective analysis of the international political circumstances can justify the military rampage in Indochina. No increase in security has been won by the raising of the ante, year after year, in an arms race already run past the finish line. It is as if the repeal of the historic premises of inequity were to be answered by an eruption of violence equal to the cumulative sum of that which held the great mass of mankind in subjection throughout history down to this day. It may be, in the words of Lincoln, that "all the wealth piled up by the bondsman's unrequited toil shall be sunk and every drop of blood drawn with the lash shall be paid with another drawn by the sword."

Citizens who ought to know better, including many of the young, have been reaching for the same alibi. Science, they say, is amoral and so tramples over human values. Technology, the offspring of science, obeys its own internal imperatives in pursuit of infinite innovation, without regard to utility or peril; its giant, intricate enterprises have no place in them for individual human conscience and responsibility. Men will do well to serve as cogs in the machine until the ultimate machine dispenses with them. Civilization, according to the new millenarians, is fated to end if not by thermonuclear war then by suffocation on the neo-Malthusian plan that says "population is pollution."

This book—composed of essays addressed to various issues on diverse occasions over the past decade—argues against such fatalistic alibiing. Just as a scientist can accept no authority for truth or error but his own judgment, so each man has personal responsibility for his role in the civilization of high technology. The same pragmatic ethic makes fulfillment of the individual human being the test of all social action. Science and technology increase men's freedom because they enlarge the choice not "merely" of means but of ends, for means always inexorably comprehend ends. Only free men—men as free as their understanding allows them to be—can make the choice of life against death. And only a self-governing society of men who have made that choice can turn the power of science and technology to extend the opportunity of a fully human existence to all of its members.

Just why, in each of several cases in recent years, American society could not have expended equivalent initiative, treasure, and talent on the exploitation of technology for more productive purposes I have attempted to say elsewhere in these pages. Here let the explanation be supplied by the accleration of history. Objective knowledge is accumulative. Its accumulation tends to proceed at an exponential, that is, accelerating, rate. This is because new knowledge, like a new biological species, creates niches to be filled by new knowledge. The last two doublings of the stock of objective knowledge have occurred, by conservative estimate, since 1900. Man has acquired most of his present technical virtuosity, therefore, only recently.

The institutions of Western society, evolving more slowly against the friction of custom and vested interest, are adjusted to managing a technology of lesser competence. They were not designed to realize the generous possibilities of largely automatic production and its abundance; on the contrary, the advance of technology has now terrifyingly capacitated the inequity and cruelty built into those institutions from the beginning of history. To charge science with the abuse of its values and technology with the contemporary increase in the incidence of man's inhumanity to man is to reach, perhaps, for proximate causes. Surely, however, the right target for constructive protest and revolutionary zeal is the web of obsolescent institutions and values carried over from the past.

The acceleration of history recurs as a theme throughout this book. This and other themes bring the essays together in eight parts, in which they appear in chronological order. The essays are presented as they were delivered to the audiences, principally academic, for which they were prepared. I have made elisions to spare the reader, somewhat, my habit of repeating myself, but my hindsights are confined to this preface and the notes preceding each chapter. The book thus sets out my efforts to make sense of the events of the past decade as they transpired and, once again, now.

The plan of the book, as suggested by the section titles, goes as follows:

Under the title "Common Sense," which is meant to echo Tom Paine, I address a deficiency of nerve and logic found at Colby College in 1962 and at Harvard University in 1967. From such deficiency comes a contemporary revival of the discredited notion that there are two ways of knowing, one to teach "how" and the other "why." The scientific revolution has exposed the fallacy that distinguishes the discovery of ends from the fashioning of means—the dualism that has so often ratified the use of bad means for claimed good ends. Scientists as well as humanists have shirked their revolutionary duty in this period. National policy in science and in the uses of science reflects the want of moral leadership in the men and institutions that should have supplied it.

The essay "On Recollection," written to defuse anxiety about the computer, deals incidentally with another popular fallacy about science. In the "knowledge explosion" people have found confirmation of the notion that the work of science is the collecting of facts. The library, the external memory of our species, supplies the metaphor to show that the connection between facts—especially the unexpected and suppressed connection—is the true object of rational inquiry.

In the next part I ask why our democratic institutions have not used to happier effect the power of technology to banish misery and ignorance from the land. Here the reader will find reference to the polluted air, the blighted landscape, and the shoddy goods, since more widely advertised by the "consumerism" and "ecology" movements. From the same symptoms I come to a prescription different from those now popularly urged. I do not find that the paved-over countryside, urban sprawl, and an overweaponed military establishment are the natural creations of high technology. These untoward consequences, I maintain, follow in part from commitment of this capability to management by the *sauve qui peut* morality prevailing in the business system and in equal part from the immobilization of the electors by the folklore of our economic institutions that still holds them in thrall. To hear the voice of the people, at least as it is relayed through the unstable amplifier of our mass-communications system, one would not guess that more than half the American people (as specified in the last essay in the book) now depend for their livings on money laid out from the public treasury by political decision.

Yet there are cheering signs of acceleration in our social and political evolution. The resilience of our democratic institutions is demonstrated by the crossing of that watershed in the redistribution of incomes to 50 percent of the people; the public payroll and public outlays have taken up most of the slack of disemployment by technology. Another watershed is being crossed with the popular acceptance of the right to a decent living. As recently as 1964, when I mentioned it aloud in public, the guaranteed annual income was heresy. Now it has been floated as a federal undertaking by the Nixon administration.

Though blurred in the ambiguity of welfare reforms proposed by the administration, the underlying ethical question has now been framed as a political issue. The enactment of the right to a decent living awaits the next swing to benevolence in our politics.

The right to health, as another ethical dispensation of abundance, calls upon a wide spectrum of technology beyond the narrow band of medical care: the topic of the fourth part. This country boasts the world's most extravagant and inhumane medical establishment. There is a giddy insight into prevailing values in the fact that Americans spend more on medicine than on education. Yet, not only the poor who are denied access to medical care, but the rich who enjoy the best of it, discover that tender, loving care cannot be bought.

Questions of ethics have high visibility in medicine. The mismatch between the public need and the capabilities of the apparatus assembled in response to the demand of paying customers can be seen not only in the clinic but just as plainly in the course offerings of the medical schools and in the motivation of medical research. Open-heart surgery and the artificial kidney have held priority ahead of the pharmacology of heroin and the labile genetics of the gonococcus. Now that public funds are articulating the needs of the poor, metropolitan medicine is finding its way in alongside country-club medicine.

The overall improvement in people's material circumstances has had more to do with the lengthening of life expectancy in the United States than the science and practice of medicine. Yet, even though the ills of the poor must await the cure of poverty, easier access to medical care can be decisive in the lives of the most needful: the infant and the aged poor. The country's medical technology, and the economics that distributes its benefits, present in scale model—admittedly in large scale—the issues that beset the harvesting of general welfare in this country from technology in general.

As foreign relations must be the extension of domestic politics, so the mismanagement of technology at home must be reflected in the way the country carries weight abroad. Hindsight shows I was most naïvely wrong about the prospects for significant U.S. technical and economic assistance to the developing countries. In spite of decline in appropriations by Congress, and within earshot of Tonkin Gulf, I was so wishful as to think that the U.S. business system would discover here a big-ticket, single-customer market as attractive as that furnished by the Pentagon.

My arguments in support of this expectation, which I believe still to be cogent, are set out in the fifth section. It will be seen that I was not relying upon the good will I know my fellow citizens so generously possess. Even so, it has been easier for the most well-bred people in America to advise India that it has too many Indians than to extend the resources, even at a profit, necessary to help the Indians and others on to the enterprise of industrial revolution. Self-interest, in this tragic debacle, could not overbalance the weighty homilies of folklore.

The denial of "foreign aid" is, of course, deeply connected to the pathology that lays out, instead, more than \$35 billion in a single year to devastate lands and peoples that looked to the United States for aid. Under Harry S. Truman's Point Four the United States was to have contributed \$2 billion to an annual \$6-billion capital input from the rich world to the poor. In thirty-five years, it was calculated, such external investment would have triggered self-sustained growth in the preindustrial nations. The wanton, bullying war in Vietnam has just entered its second decade.

There is little in the record since World War II, therefore, to suggest that American foreign policy is illuminated with any comprehension of the change in the condition of man portended by the industrial-scientific revolution. On the contrary, the enormous appetite of the American economy has been sustained at the expense of the peoples labeled in cold war terminology as "Third World"—by the rape of their resources, by the stagnation of their economies, and by the frustration and even subversion of progressive leadership wherever it has raised its head within the reach of this country's power. The celebrated economic growth of South Korea and Taiwan is the byproduct of lavish military subsidy of military regimes. These adornments of policy are reduced to real-life dimensions when viewed against the mournful background of South Asia, Africa, and Latin America. The industrial revolution is gathering momentum there despite United States policy, but at greater human cost because of it.

To putative friends as well as foes, the United States appears today as a nation missing its head as well as lacking in heart. This is a perceptive diagnosis. The institutions that should have supplied intelligence and conscience to the making of national policy have taken leave from their constitutional function throughout the twenty-five years since World War II. The universities scarcely demobilized at the end of the war; the great ones, metamorphosed to "multiversities," made "service" a function coordinate with teaching and research. They operated major arsenals for the military, fronted for the intelligence agencies, and conducted research on exotic weapons and modes of warfare. In return the federal government, with military agencies supplying much more than half the funds, financed 80 percent and more of their research budgets.

University administrators and leading academic entrepreneurs have argued the propriety of their arrangements with the federal government. The independence and purity of motive held by the faculties were not compromised, they say, by the grants and contracts for research conducted in the "universities proper"— a new entry in the federal ledger that distinguishes them from the research institutes and facilities they operate for the government. The same interests testify also to the high regard for academic freedom displayed by the federal funding and contracting agencies. In justification of these arrangements, they point to the extraordinary bloom in scientific achievement that has put the United States ahead of all countries in the world.

The ultimate weapons are now emplaced, however, and the last major space adventure, for a while, has been engineered. All at once the federal money has stopped. The multiversities share the depths of the economic recession with other military contractors. They find it difficult to win public sympathy because the public never rightly understood that money appropriated for the development of weapons was going to the support of pure science and higher education. To the extent that science and the university are now identified with the ugly business of war, the case for their support, never made to the public, is now more difficult to make. An informed public wonders whether technologies of interest to the military have not been given undue emphasis and whether the country has an adequate engineering force trained in humbler disciplines required for operation of the national household. Uncertain telephone service and power black- and brown-outs in many parts of the country have not been inspiring confidence. The U.S. heavy machine-building and electrical industries have been losing customers even at home to foreign competition on quality as well as price of products. A corresponding misplacement of emphasis in the life sciences has been referred to above.

In the political and social sciences, relations with Washington have worked more easily recognizable corruption. The chairs in these departments are occupied by "in-and-outers," who shuttle between appointments in the universities and in Washington. In the university, their prestige derives from their connections to power and their access to privileged archives. To Washington or to the political faction that brought them there, they return a steady positive feedback of rationalization for official policy in whichever field they are implicated.

In foreign affairs these scholarly labors have now been thoroughly discredited. Washington is the last capital to cling to the simplistic cold-war model of the world polity. Allies and enemies of the United States are finding their way into multilateral relationships with one another and with powers emerging among the developing countries. The United States faces a new isolation imposed from without.

Similarly, in domestic affairs, timid fiscal prescriptions from the New Economics have proved powerless to restrain inflation or reverse recession and have been confounded by the running on of both together. Few occupants of the distinguished university chairs have shown they are aware of the obsolescence of the premises and values of our economic institutions at the present culmination of the technological revolution.

This is the treason of the clerks. With notable exceptions, they have not provided the independent intelligence and criticism necessary to wisdom in public policy and surveillance of policy by the public. An example of the fortitude to be expected of the scholar is provided by those physicists who have ventured into the political forum on the issue of arms control and disarmament. They made the weapons and know them more intimately than the military and political authorities. On the hazardous task of getting the genie back in the bottle they have the best technical advice to offer. These scientists have pressed their advice through the channels of closed politics. And they have supplied to public debate the authoritative information and cool judgment that are the antidotes to panic and despair.

The multiversity, in the present reversal of its fortunes, is proving not much more substantial than other phantoms of public relations. The university was there behind the false front all this time. The term "university," as used here, is like democracy: reference is to the ideal. That ideal is a community of scholars, organized for the increase and diffusion of human understanding. By derivation from the sovereignty of the citizen, it is hedged with the immunities that place his conscience and conviction beyond reach of other authority. In its autonomy, the university is a self-governing community; here consent to reason is the law and coercion has no place.

The university is thus a moral community. It thrives to the degree that each member makes good his commitment to the common cause. As a center of innovation in technique and purpose, the university is the object of fear and wonder in the larger, surrounding community. The learning it fosters may at one time invite repression from outside and at another time exploitation. For all its apparent prestige and real power, however, the university is a vulnerable institution. It is vulnerable especially from within, to default by any member on the moral commitment that binds the community together.

That is why many universities fell so ignominiously to the first challenge laid by the student rebellion. Prior breach of the compact by senior members had left these communities without moral authority over their junior members. When the dust settled and the police departed, there was still no honorable response to the hard questions put by the students but concession.

Out of turmoil, the universities are putting themselves to-gether again as communities of scholars. Trustees, faculties, and students are joined in formal and informal constitutional con-ventions. They are working out new modes of "governance," formulating the ethics of relations with powers outside, and overhauling the curriculum, with regard being paid to the relation between teachers and students as well as to content. The college is returning as the center of the university. This is a decisive development, for it is the undergraduate's innocent preoccupation with meaning, connection, and value—the pres-ent cry for "relevance"—that forces concern for the unity of knowledge in opposition to the divergent trending of the disci-plines into comfortable isolation from one another in the graduate schools and departments.

In the reinvention of the American university, now so hope-fully proceeding, a paradox remains to be resolved. Stating the paradox as a question: How can the university secure its auton-omy and, at the same time, its support by public funds? The great private universities of the Northeast are less well prepared to show the way than the state universities elsewhere in the country. Experience with federal funding over the past twenty-five years has shown that these institutions, whether state or private, cannot accept support in compensation for services rendered. To the extent that they have served as "instruments of national purpose" they have failed their antecedent obliga-tion to the citizenry. The student rebellion, along with the phasing-out of federal research programs, has recalled the uni-versities to their primary mission. This is to foster the work of free inquiry by which sovereign citizens frame the national along with other purposes. Only in the fulfillment of that mission can the universities claim their autonomy and their right to public support. If this seems an intricate argument to make to the taxpayer, that is a measure of his need for educa-tion in self-government.

In the last section of the book, I have tried to look down the way less plainly marked from the fork in the road. My object is not so much to see far as it is to find alternatives in what is near at hand in the present. I am preoccupied by the state of our own nation because the United States, if no longer celebrated as the world's best hope, still strongly engages the hopes of the world. To my hearers, my fellow citizens, I argue that none of us can any longer delegate his personal responsibility to the absolute of national defense and the mysterious workings of the market's unseen hand.

The military forces of the United States have acquired their excess of overkill largely by default. The public and private interests in armament have built the machine without significant opposition and, in the main, without public understanding of what they were doing. No competing interest advanced a comparably effective claim upon the human and material resources that were squandered upon arms, even though there existed at the beginning of this decade a well-advertised, long-standing inventory of aching domestic needs—an inventory that has lengthened with the years. Economically and politically, it was more expedient to put those resources down the sink of weaponry.

Expenditures on armaments, on para-military exploits in high technology, and on shooting war have aggregated 10 percent of the gross national product since the end of World War II. Sustaining consumer demand in the domestic economy, these expenditures have pushed economic growth through the country's longest business boom. The rate of growth has not been high compared to the record of other, better-managed economies. Some of the potential has been bled off by the venturing overseas of the great U.S., now "international," corporations. (In the European business community it is asked: In whose interest are these corporations managed—their managements'?) Growth has nonetheless carried the rate of consumption so high that the attainment of a comparable standard of living for the rest of mankind would seem to lay a forbidding demand upon the earth's finite resources. Yet such abundance has not ended poverty in this country.

The continuation of economic growth has meanwhile relieved Americans of the obligation to face the question of equity—the promise to all citizens, in their social compact, of equality of opportunity. So long as most people sensed rising well-being, they could believe that sheer growth was meeting that obligation, automatically. Growth, moreover, propagated the Puritan ethic to numerous new members of the middle class, whose ethnic traditions had formerly taught them to rely upon God's mercy. Once more reaffirmed in the popular view, and in the works of such scholars as Daniel P. Moynihan and Edward C. Banfield, was the holding that poverty is the outward sign of inward lack of grace.

Now that growth has stopped, ominous tensions divide American society and cruelly isolate its impoverished minority. The division is sharpened by the association of poverty with race; the isolation is deepened by the metropolitan segregation of the non-white poor. No spontaneous process promises cure for these ills, which trace their beginnings to the white man's first settlements in America. Growth has failed.

Growth has failed even those who enjoy its bounty. Suburbia —what Clyde Kluckhohn called "the climax of American civilization"—is now the "outer city." The planless dispersal of the city has not only dissipated the values of urban life but has also proved to be the most insidious pollution. Conservation and zero population growth have been ratified as new amendments to the American creed.

My fellow citizens are ready, I believe, to consider the advantages of the stationary state; that is: the economy that aims at equity rather than growth. As they approach this revolutionary adjustment of their values they will find that the acceleration of history has abolished many old issues and transformed others. To cure unemployment, for example, does not require further increase in the surfeit of material goods that already burdens the market apparatus. Automatic production has already spun the majority of the labor force out of production jobs and out of the production sectors. The "services" are the fastest growing sectors and government services the fastest growing of these. Typically, the latter are those that benefit the entire community and not the individual consumer alone. It takes deliberate, rational, political decision to deploy demand for these. Conversely, demand is rising in the community for those same social, if not economic, goods—education, the public common of the wilderness and of the urban ambience, health services, and the like—that issue from political action. Wonderful to say, these goods are not diminished but are increased by the enjoyment of them.

Generally speaking, there is less opportunity for people to improve their circumstances at others' expense. The crucial decisions and actions to be taken in the interest of self and one's posterity are those in behalf of the community.

I The Acceleration
of History

The Acceleration
of History

ACCELERATION does not characterize the evolution of the universe until that process gives rise to life. Deceleration seems to be the rule that prevails in inanimate nature. According to the late George Gamow, elaborator of the evolutionary or "big bang" cosmology, the first major event in the history of the universe—the generation of the elements from the primordial gas of energy—occurred within 30 minutes after moment zero. The next milestone—the transition from the regime of energy to that of matter—comes 250 million years later. There follows a longer stretch of time before the universal haze breaks up into clouds that are to condense into galaxies, and an eon yet before the first stars ignite in the galaxies.

The universe had attained an age now estimated at 10 billion years, and the earth had been turning for 1.5 billion years in the sunlight when life appeared on this planet. With that event, only 3.0 billion years ago, nature's clock began to race, at least in this corner of the expanding, cooling, darkening universe. Some benchmarks of the shortening of the time intervals in the evolution of life to higher and more complex forms are cited in the essay entitled "Last Words from the Dinosaurs."

Life is so entirely improbable in this and other respects that its contemplation can send elderly physicists, even today, to searching for relief in metaphysics. The perfectly satisfactory natural explanation need not detain this discussion. My purpose here is to consider the most recent and improbable of all events in evolution: the invention of value.

This is an event quite plainly identified with the evolution of man. The term "invention" is preemptive; its inexplicit premise excludes the notion of value as an absolute to be disclosed by revelation or discovered in a syllogism. To state the point in the provocative, I say: "Science is the ultimate source of value in the life of mankind." In the essay that follows I go on to show how values have evolved in correspondence to the accumulative progress of science and technology down to our day. It has now become possible to cherish every human life, and so it is no longer possible to justify the spending of some human lives for the ecstasy, pleasure, profit, satisfaction, enlargement, benefit, or higher purpose of others.

Such historical-pragmatic account of value does not carry the clean generality found in absolute statements. The philosophy of science can, however, meet this humane, aesthetic need. Consider how the molecular biologist Jacques Monod put the matter to his brothers upon his accession in 1970 to his chair in the Collège de France. With the puritan astringency of his Huguenot forebears, he said:

In the ethic of knowledge the single goal, the supreme value, the "sovereign good" is . . . objective knowledge itself. . . . There must be no hiding that this is a severe and constraining ethic which, though respecting man as the sustainer of knowledge, defines a value superior to man. It is a conquering ethic . . . since its core is a will to power: but power only in the [realm of ideas and knowledge]. An ethic which will therefore teach scorn for violence and temporal domination. An ethic of personal and political liberty; for to contest, to criticize, to question constantly are not only rights therein, but a duty. A social ethic, for objective knowledge cannot be established as such elsewhere than within a community which recognizes its norms.

Henry Adams, who supplied the theme of this essay, supplied as well its countertheme. It may be that the Virgin has gained more confessants now among the young than the Dynamo in 1963, when I wrote this essay for the students at Phillips Academy, Andover. The material abundance of our land has become a source of regret and remorse to its beneficiaries. Alienation from the values of American middle-class culture has been accompanied by alienation from science. It is, in fact, the recoil of the young that prompted me to attempt this postscript to the essay.

Material abundance is not of itself a good. It does, however, offer the option of good. That a society not ready for the choice may misuse and abuse wealth won from nature does not impeach the option. It is the new choices, opened up to the species by the increase in its knowledge and power, that have forced the evolution of human value from age to age. By this route my argument reaches its point that science is of itself a good. Science is not, as its detractors claim and so many defenders concede, value-free. It is an enterprise that teaches ethics in its practice, as Monod has shown, and in its yield of deeper understanding and power as well.

Weak, inadequate, and even cruel and crooked men may practice science. It is in their best hours and to its rigorous standards that they practice, however, or they fail. Modern technology may have its barbarians as well, but their crimes must come to judgment under higher law than that under which history arraigns their predecessors. The evolutionary destiny of our species requires, from generation to generation, that all men know better.

The occasion for this essay was the dedication of a magnificent new building for the teaching of science at Andover. The invitation to make this contribution to the celebration brought me back to the school for my first visit after thirty years. I had been one of the students who had the good luck to be befriended by Dirk H. van der Stucken. A man of great learning and generosity, Van der Stucken taught us Latin and German in the classroom and gave us general instruction in civilization during unstinted hours of his extracurricular time. My memory of the school, without him, recalls a big, remote, uncompromising community in which one did well to swim and not to sink.

The Andover to which I came back was a very different school. It was lucky I had been inspired to illustrate my lecture. For, along with science, the arts had commanded the principal new addition to the school's resources, gathered in under the leadership of John M. Kemper, its headmaster from 1948. The Addison Gallery of American Art, which so nicely balances the Oliver Wendell Holmes Library on the symmetrical Georgian campus, had been a nice place to visit. Now this gallery is tied integrally to the educational process by studios, workshops, projection rooms, and a theater, all resounding with enterprise and demonstrating that the mind grows by action and emotion and not by the book alone. What the faculty calls "affective" learning is no longer left to the chance encounter of student and teacher. It claims as much time as "cognitive" learning on the school's agenda.

There is more about Andover in this book. This return to my prep school was the beginning of a new involvement in it and the vital phase of education in which it is engaged.

T HE sciences are as natural to a liberal education as the humanities. The truth-seeking and toolmaking enterprise of rational inquiry does not encompass all of the concerns of man. Remote recesses of the interior experience of life and distant reaches of the exterior universe will remain always to lure and defy the inquirer. Today, however, there are few realms of human concern to which science is not relevant. An understanding of science is essential to each man's orientation in the world as it is known to the mind of man. Both as knowledge and as process, science responds to the troubling questions: Who are we? Whence have we come? Whither do we go? Moreover, science shows us how to ask such questions in a more productive form. Beneath the surfaces of things accessible to the unaided senses, science discloses forces, dynamics and transformations, symmetries and diversities, order, precision, and grandeur unknown to previous generations.

These are some of the considerations—philosophical, moral, and aesthetic—that motivate the present increase in emphasis on science in the curriculum of Phillips Academy and other schools.

It was a cowardly and costly truce in the academies of the nineteenth century, at the close of the great Darwinian scandal, that set up the false dichotomy between the sciences and the humanities—as if truth could be sought in the absence of concern for value or value cherished without courage to face the truth. The breach is now being closed, perhaps more quickly in our secondary schools than in the colleges.

There is still another consideration, however, that compels emphasis on science in our schools. An understanding of science becomes increasingly essential to the exercise of citizenship in a civilization that is being transformed by science. As long as thirty years ago, when I was here at school, people were saying that science had outrun society, that man's control over nature had outstripped his capacity to manage himself. The peril we live in now—the shameful catastrophe that overhangs civilization—places each of us under responsibility we cannot delegate to others.

Our peril was prophesied at the turn of the century by one of the few historians who has ever undertaken to explore the relations between science and the history of society. Henry Adams knew something of the work of Faraday and Maxwell in electromagnetism; he comprehended the main theme of the thermodynamics of Josiah Willard Gibbs; in his private speculations, he attempted to deal with the dilemma in which physics was placed by Michelson and Morley, and he was ready for Becquerel's discovery of radioactivity. Looking back over the tumult of modern history, he plotted the rising curves of the rate of scientific discovery, of coal output, of steam power, of the transition from mechanical to electrical power. Any schoolboy, he said, could plot such curves and see that "arithmetical ratios were useless"; the curves followed "the old familiar law of squares." That is, they rose more steeply as they ascended from the time base line. The logarithmic scale of the time base line of Adams's chart gives equal space to the last millennium and to the preceding 10,000 years. It thus not only serves geometrical

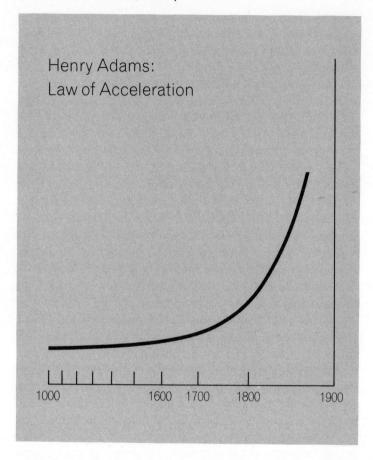

Henry Adams:
Law of Acceleration

1000   1600 1700 1800   1900

convenience but also reflects the compression of the past in our memory.

"The acceleration of the seventeenth century," Adams observed, "was rapid, and that of the eighteenth was startling. The acceleration even became measurable, for it took the form of utilizing heat as force, through the steam engine, and this addition of power was measurable in the coal output." Acceleration was the law of history.

From his acquaintance with Gibbs's historic work, Adams was prepared to recognize that such changes in quantity amounted to changes in quality—"changes in phase" in the language of thermodynamics. The history of thought was in passage through three phases, each with a duration in years that was the inverse square of the duration of the preceding phase.

Supposing the Mechanical Phase to have lasted 300 years, from 1600 to 1900, the next or Electric Phase would have a life equal to √300, or about seventeen years and a half, when—that is, in 1917—it would pass into another or Ethereal Phase, which, for half a century, science has been promising, and which would last only √17.5, or about four years, and bring Thought to the limit of its possibilities in the year 1921. It may well be! Nothing whatever is beyond the range of possibility; but even if the life of the previous phase, 1600–1900, were extended another hundred years,

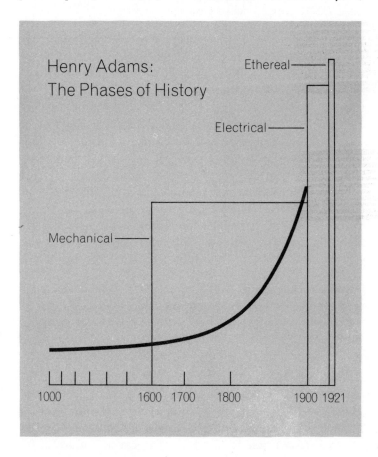

the difference to the last term of the series would be negligible. In that case, the Ethereal Phase would last till about 2025.

By Adams's calculation our lives are on probation: we are living in the grace period between 1921 and 2025. Adams despaired of our capacity to withstand the overriding force of acceleration. In 1905, he declared,

Yet it is quite sure, according to my score of ratios and curves, that, at the accelerated rate of progression since 1600, it will not need another century or half century to turn Thought upside down. Law, in that case, would disappear . . . and give place to force. Morality would become police. Explosives would reach cosmic violence. Disintegration would overcome integration.

He likened the life of thought to the passage of a comet: "If not a Thought, the comet is a sort of brother of Thought, an early condensation of the ether itself, as the human mind may be another, traversing the infinite without origin or end, and attracted by a sudden object of curiosity that lies by chance near its path. If the calculated curve of deflection of Thought in 1600–1900 were put on that of the planet, it would show that man's evolution had passed perihelion, and that his movement was already retrograde."

The same nightmare vision of the comet has recurred more recently to another man. That man, of all men, was H. G. Wells. In the last year of his life, within twelve months after the first public demonstration of the new explosives of cosmic violence, he wrote:

Events now follow one another in an entirely untrustworthy sequence. . . . Spread out and examine the pattern of events and you will find yourself face to face with a new scheme of being, hitherto unimaginable by the human mind. This new cold glare mocks and dazzles the human intelligence . . . no matter how this intelligence under its cold urgency contrives to seek some way out or round or through the impasse. . . . The writer has come to believe that the congruence of mind, which man has attributed to the secular process, is not really there at all. . . . The two processes have run parallel for what we call Eternity and now abruptly they swing off at a tangent from one another—just as a comet at its

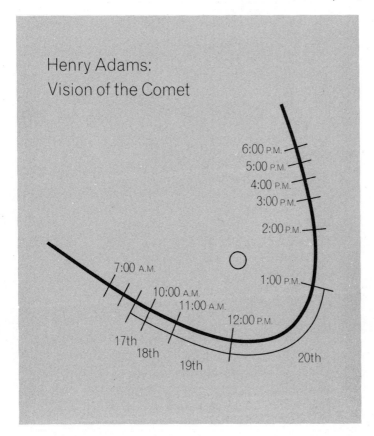

Henry Adams:
Vision of the Comet

perihelion hangs portentous in the heavens for a season and then rushes away for ages or for ever.

Most of us are less well prepared than Adams or Wells to explore the relevance of science to history. Yet we may charge to their personal circumstances the despair to which each man came at the end of his studies; for each had come as well to the end of life. I am not here to urge their despair upon you.

On the contrary, it is my thesis that the dilemma of our age—the catatonic indecision of men and nations in their present perilous confrontation with the choice of life or death—follows directly from the general failure to understand that science is relevant to history. That failure is compounded by another: the failure to comprehend the lesson that is pointed by the relevance of science to history.

I shall state the lesson in didactic terms in the hope to state it plainly: Science is the ultimate source of value in the life of mankind.

Man's ascending mastery over the forces of nature has progressively transformed not only the relationship of man to nature, but the relationship of man to man. From age to age, discovery and invention have opened new scope and possibility to human life. With each new possibility has come the necessity to choose. In the succession of choices, the moral and the social order—men's ideas of good and evil and the institutions that embody them—have evolved. The now steeply accelerating advance of science allows no time for evolution. We are compelled to an immediate reexamination and deliberate overhaul of the values and institutions that we have carried into the present from the swiftly receding past.

Let us spread out the pattern of events and see how it sustains this thesis. I have projected Henry Adams's curve backward to the classical period, 2,000 years ago, and forward 50 years to the present. A wealth of statistical data supports the ascent of the curve to twice the height at which Adams charted it at the turn of the century. Every index, from the consumption of energy *per caput* in the United States to the volume of scientific publication, has more than doubled in this period. Since few statistical series reach beyond 150 or 200 years into the past, the projection of the curve back behind 1600 must be regarded as largely symbolic. The projection is documented, however, by a number of crucial indexes, especially if one may reverse the Adams rule of historical phase and count changes in quality as accumulations in quantity.

Consider, for example, the shrinking time intervals from the discovery of one primary force of nature to the next. At the outset, there had to be the notion of a natural or inanimate force, as distinguished from the animistic and particular *genius* of the place, thing, or process; this elementary idea was first propounded in the science of the Greeks two millennia ago. Then, 300 years ago, came the Galilean-Newtonian great world

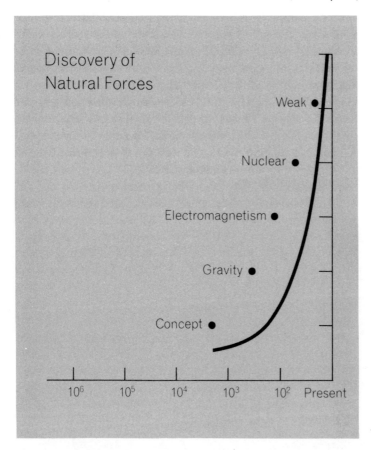

Discovery of
Natural Forces

Weak ●

Nuclear ●

Electromagnetism ●

Gravity ●

Concept ●

10^6 10^5 10^4 10^3 10^2 Present

system ordered by the force of gravity. Next, only 100 years ago, comes the Franklin-Faraday-Maxwell discovery of the electromagnetic force. Now, only 50 years ago, Einstein, Planck, Rutherford, and Bohr uncover the most energetic of the forces, that which binds the nucleus of the atom. Last, at this very moment, physics is comprehending the presence of the fourth primary force of the universe: the so-called weak force observed in the decay of elementary particles.

Consider, alternatively, the curve that is plotted by the successive isolations of the 92 elements into which matter is chemically differentiated. Perhaps as many as a half-dozen elements—carbon, copper, gold, silver, iron—had come into use in more or less pure form at the dawn of recorded history 10,000 years ago. The number rose abruptly to 20 with the beginning of modern chemistry 200 years ago. By the time Mendeleev laid out the table of elements, a century ago, the number of elements isolated had more than doubled again to 60-odd. Today physics has carried the series out beyond the bounty of nature, adding 10 so-called synthetic elements to the table—and the curve might be extended indefinitely by grafting on the lengthening table of fundamental particles.

Consider still another curve—that plotted by the mastery of the major sources of inanimate energy. The starting point

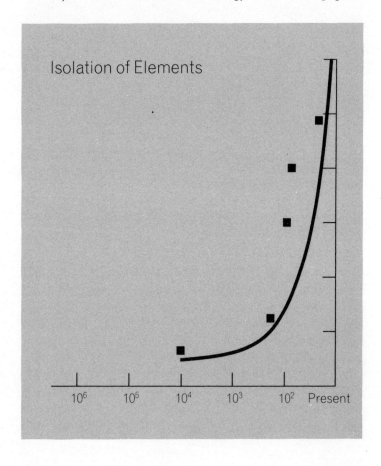

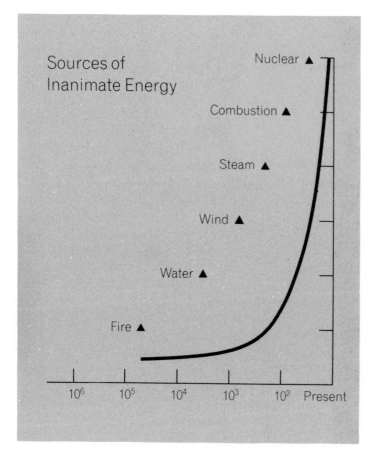

carries our projection back behind the beginning of history to at least 50,000 years ago when man discovered the first uses of fire. With his own vital energy amplified by fire, man must already, in that remote time, be reckoned as a geologic force. He used fire not only to warm his body and to cook his food but more significantly to burn forests and extend the grasslands over which he could hunt more safely and productively. The next point on the curve marks the harnessing of water power in the Bronze Age, 5,000 years ago. It was only 800 years ago that man began to make comparable use of the wind; the invention of the windmill in twelfth-century Europe was something of a

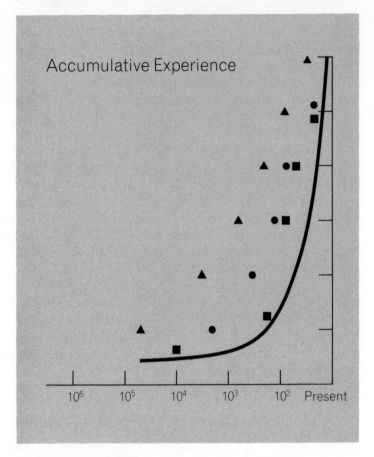

technological revolution, spreading in a century from Normandy to the Black Sea. The industrial revolution itself began, of course, with the harnessing of steam, 200 years ago. Direct or internal combustion—which generates as much mechanical and electrical energy as steam in contemporary technology—dates back only a century. And now the first nuclear reactors are delivering electricity to the power networks of the world.

The pattern of events shows history on a course of accelerating acceleration. The major developments in man's accumulative experience have occurred within the most recent times, and these developments occur at shorter time intervals into the very present. This chart does not, of course, begin to tell the full story of history. Plotting only the accumulative elements of human experience, it excludes the rest—the glory and tragedy, the shame and honor, the bestial and the humane—and so, by some lights, it excludes all that gives meaning to history. Yet the exponential curve of science, I believe, plots the mainstream of history insofar as history has not merely repeated itself. This assertion is confirmed when we place the brief period of recorded time in the perspective of man's longer past. At the giddy vantage point on the vertical coordinate to which we have ascended in our lifetime, we are as far removed from the nineteenth century as our grandfathers were from prehistoric time.

The starting point of the plot of history may now be established as far back as 1.7 million years before the present and proportionately closer to the time base line. At a site reliably dated to that distant time, on a buried lakeshore in the Rift country of Africa, anthropologists have recently unearthed an assemblage of stone tools. With these tools they found fragments of the bones of the hands that had made them. The hands are not human hands—not our hands. They are the hands of a primate who still used them at times for walking. In the old taxonomy of primates it was supposed that man had made the first tools; toolmaking was the status symbol of membership in our species. Now, it would appear, tools made man. Certainly, toolmaking conferred a competitive advantage on the maker of better tools. But the meaning of this phase of history goes deeper. The truth is: man made himself.

The record as to bones of hands and skulls is scanty. There is an abundance, however, of the fossils of behavior—the stone tools. In their increasing diversity, specialization, and refinement, they give evidence of the evolution of the hand and of the brain, of which the hand is an extension. The tools show, in time, that evolution has quickened because it has entered on a new mode. It has become cultural as well as biological— Lamarckian as well as Darwinian in that acquired characteristics are transmitted from generation to generation by teaching and learning. Emergent man has already discovered in his own head

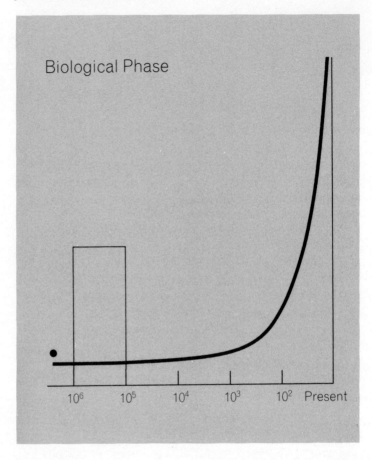

the notion of purpose, for which men have since sought valida-
tion in so many other corners of the universe. The increasing
specialization of the stone tools implies, of course, a corresponding elaboration of technology employing less enduring materials. The mastery of new environments commanded thereby
disclosed new possibilities and new ways of life—one might go
so far as to say new goals and values—to the men who sired
modern man.

Although the earliest bones of *Homo sapiens* are dated only
25,000 years before the present, indirect evidence places the
origin of our species at a date no more recent than 100,000 years

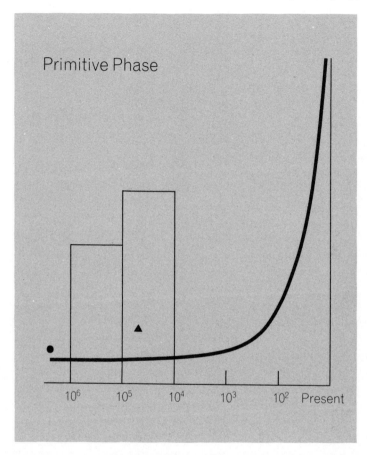

ago. As long as 50,000 years ago, the diversity of hunting and
food-gathering technologies enabled man to make himself at
home in every environment on earth. Some primitive cultures
have persisted even into modern times on the Arctic shores of
the northern continents, in the interior of the southern continents, and on oceanic islands in the Pacific. These peoples have
taught us to use the term "primitive" with respect. There is no
human language that is primitive; each has a grammar as well as
a vocabulary. The aesthetics of these cultures is the more compelling because it so directly articulates the experience of life.
Typically, their social order is the extended family and the code

of law and custom submerges the individual in the common identity and destiny of the group.

It has been shown that even in a continental environment as favorable as that of North America, primitive technology could not sustain more than one person per 10 square miles of territory. The aboriginal population, of course, never approached the limiting density. This may be taken as a measure of the hazard and uncertainty of the hunting and food-gathering way of life that is the common theme of primitive religion and art. Imperceptibly, over tens of thousands of years, as certain of these peoples came into possession of more intimate understanding of their environments, they found a more secure way of life, as herdsmen and cultivators of the soil. By 10,000 years ago they had domesticated all of the plants and animals now grown on the world's farms. There could be no doubt about the progressive nature of this development. It multiplied by 100 the potential size of the population that could be sustained on the land. As history was soon to show, the labor of four families in the field could now support a fifth family in the city.

The transition to agricultural civilization was made in the same 2,000–3,000-year period in Asia Minor, in the valley of the Nile, in the Indus Valley in India and in China; the transition was made more recently in pre-Columbian America, apparently in entire independence of events in the Old World. Wherever this revolution occurred, it gave rise to essentially the same social and economic institutions. The function of these institutions was to secure the inequitable distribution of the product of the soil. Law and custom speedily legitimized the necessary measures of coercion. But the primary compulsion, as we can now see, was supplied by the slant of the curve of discovery and invention. Over millennia or centuries, progress was substantial; in the lifetime of a man, however, it brought no appreciable increase in the product of his labor. Population tended always to increase faster than production, maintaining a constant equilibrium of scarcity. Bertrand de Jouvenel has described the situation with precision: "As long as there is a fairly constant limit to production *per caput*, one man can gain wealth only by making use of another man's labor; only a few members of society can gain wealth, therefore, and at the expense of the rest. All ancient civilizations rested upon the inexplicit premise that the productivity of labor is constant."

The inexplicit premise of scarcity is stated plainly enough in the plan of the ancient cities. Invariably it shows the palace, the temple, and the garrison within the ruin of the walls and, outside, the traces in the soil of the hovels of the slaves. Thus four-fifths of the population was made to render up the surplus necessary to sustain one-fifth in the new enterprises of high civilization.

For the still smaller fraction of the population who were the ultimate beneficiaries of these arrangements, it can be said that

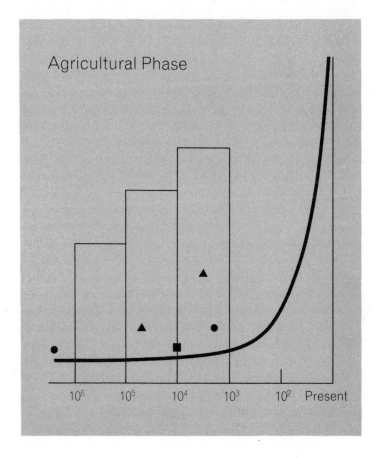

they made the most of the possibilities, opportunities, and aspirations now open to them. History is first and foremost the story of their extraordinary lives. They gave impetus to the advance of invention and discovery; they explored new realms of aesthetic experience, enriching the lives of all men with their vitality and sensibility; they pondered deeply on questions of justice and equity, setting their reflections down in codes of law and treatises on moral philosophy; they developed the arts of government and administration; they led their countrymen on bold adventures in politics, war, and conquest.

Of the 80 percent of the population who were excluded from history—except when war loosed murder, famine, and pestilence on the land—history has little to say. It has even less to say about the underlying inequity of the social and economic institutions—whether slavery, serfdom, taxes, rent, or interest—that laid the burden of history on their backs. Laws were passed to regulate the treatment of slaves and serfs, but the gross immorality of these institutions was never called in question until modern times—not until, that is, the inexplicit premise of scarcity itself had been overturned.

It is easy now to mark the turning point of history. But even in the seventeenth century people sensed the accelerative force of the contemporary deflection of thought. Without doubt, the most revolutionary idea in the life of man was the concept of inertia advanced in 1638 by Galileo—then already past the age of seventy and writing in secret under house arrest for the lesser heresy of advocating the Copernican revolution. Galileo's great insight comprehended at once the swinging of a pendulum and the motion of the planets on their orbits. The idea of inertia not only changed men's view of nature; it placed a primary force of nature in their hands. Within a few generations they were setting much else besides pendulums in motion.

The surplus gathered in by the institutions of scarcity found a new historic function. It became the wealth of nations to be invested in the increase of capacity to produce wealth. Though hindsight encourages us to place emphasis on the acceleration of the rate of discovery and invention in this period, we must not fail to credit the role of the institutions of political economy. In 1802, looking in satisfaction on the ascendance of Britain, then in the vanguard of the industrial revolution, Sir Humphry Davy astutely observed: "The unequal division of property and of labour, the difference of rank and condition amongst mankind, are the sources of power in civilized life, its moving causes and even its very soul."

Today, after two centuries of industrial revolution, we have come to speak of two kinds of nations: developed and underdeveloped. Some twenty-odd nations, comprising about one-third of the world's population, have joined in the industrial revolution. To one or another degree, their entire populations are entrained in the heady experience of increasing well-being.

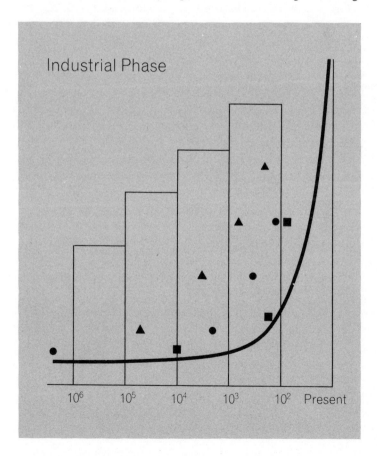

It is no accident that the nations which set out first on this course and have come furthest have also given the world the most favorable demonstration of self-government by citizenries that count no second-class citizens among their numbers. The poor nations, on the other hand, comprise a whole two-thirds of mankind still immured in the economic and social institutions of agricultural civilization and include some tens of millions in Africa and Southeast Asia caught, at this turn of history, in transition from hunting and food-gathering to settled agriculture. By all the quantitative indexes that measure the contrast between the rich and the poor nations, it is plain that history has entered on a new phase.

The signs of change in quality as well as quantity can best be seen in the life of our own country. In the United States we behold the contemporary climax of the industrial revolution. Our standard of living serves as the usual index of our revolutionary leadership. Actually this well-advertised story obscures the crux of the change. Americans, it is true, consume three times as much in goods and services *per caput* as their fellow inhabitants of industrial civilization and more than twenty times as much as the denizens of contemporary agricultural civilization. It is also true that this represents a gain of seven times over the American standard of living of 120 years ago. But the triumph fades somewhat when it is observed that our famous miracle was achieved at a rate of less than 2 percent per year per person. Contemporary industrial economies are advancing at two and three times this rate! The American celebration is dampened further when we are compelled to admit that the average hides poverty which blights the lives of a third of our people.

No, it is not our success as consumers, but rather as producers, that opens the new phase in history. Strangely, perhaps, it is the agriculture of our industrial system that most clearly exposes the nature of the change technology has brought and portends in man's way of life. In contrast with agricultural civilization, where 80 percent of the people are bound to the land, no more than 8 percent of the American labor force work on the farm. Working fewer acres each year, they produce still greater yields; presently, enough to feed 12,000 calories to each American every day—enough to feed a billion people an adequate daily ration. Suffice to say, we cannot eat all of it, and after giving and throwing a great deal of it away, we still have a surplus to keep compulsively in storage.

In the American agricultural surplus we behold a very different kind of surplus from that which was gathered by the lash in Mesopotamia 6,000 years ago. It is a true granary-bursting, physical surplus; it may be taken as symbolic of the surpluses generated elsewhere and everywhere in our industrial system. From these surpluses our economy adds to its capital at no visible cost to current consumption, maintains a vast military

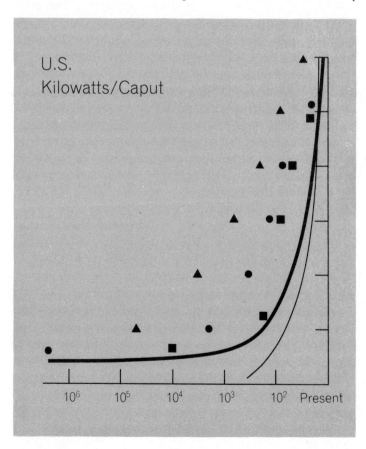

U.S.
Kilowatts/Caput

10^6 10^5 10^4 10^3 10^2 Present

establishment, and still accumulates unsold inventories that periodically throttle its channels of distribution. The industrial surplus is the opposite of the scarcity surplus: it is abundance.

America's capacity to produce abundance has long since been freed from the limitations of human muscles and nervous systems. The exponential increase in the flow of electrical energy through our economy has placed the equivalent of more than 40 human slaves (one man-year = 150 kilowatt hours) at the disposal, on the average, of every man, woman, and child. Fewer than half our labor force are employed as producers of goods; not much more than 40 percent, if the production of armaments is subtracted from the total output. Factory workers will be as scarce as farmers in another generation. Our abundance is freed equally from the constraint of resources. The industrial order does not discover resources; it literally creates them. Nor does the system waste and lie idle for lack of demand on its capacity. In addition to our private, family poverty, already reckoned here, our land presents an appalling landscape of public poverty: of blighted cities, polluted rivers, wasting natural resources, neglected schools, bankrupt health services, to mention a few of the demands outstanding.

I could go on and cite the demand for increased investment coming now from the underdeveloped nations that have begun their industrial revolutions; they will carry their revolutions forward without our help and at our greater cost in that event. I would close by wondering aloud why the extravagant demands of the space and military captains should have such exclusive claim on our surplus, especially since they can't use it all!

The constraints on the distribution of abundance, it is apparent, originate elsewhere. Inevitably, the transition to the latest phase of history, which I have called the humane phase, must confront us with a crisis in our institutions and values. The old regime of scarcity is at an end; the time has come to repeal the iron law that says one man's well-being can be increased only at the expense of his brothers. We must frame our values and institutions to respond to the new dispensation of abundance: the well-being of each man increases with increase in the well-being of all.

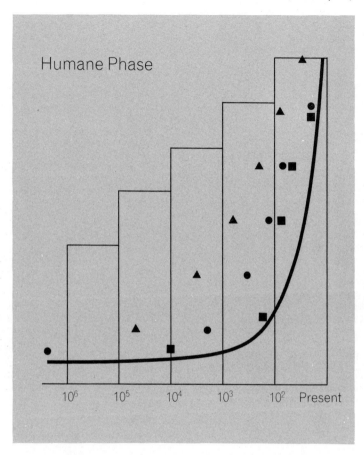

Humane Phase

10^6 10^5 10^4 10^3 10^2 Present

The menace of the power man has taken in his hands from nature has been all too well popularized. Statesmen and citizens alike have learned to reckon with it in all their calculations about the future. There are signs, in fact, that they have become used to reckoning with it. The promise of the power in our hands ought to have an equal place in the public and private consciousness. The realization of that promise must begin with the understanding that truth-seeking and toolmaking have disclosed the noblest and most generous ends to human life and have given us the means to accomplish them here on this earth.

Last Words from
the Dinosaurs

EVERY spring, usually in a city west of the Hudson River, the administration of Harvard University rallies its more active alumni for a semiacademic festival, devoted principally to the business of the alumni, which is to support their university. In May, 1963, at the meeting held that year in Cleveland, Ohio, I had the pleasant but hazardous assignment of an after-dinner speech. My theme was dictated by the holding of this dinner party in the Hall of Nature of the Natural Science Museum. The hall is dominated by a large herbivorous dinosaur paired with a smaller carnivorous dinosaur in preprandial posture.

HAPLOCANTHOSAURUS is the full name of this personage from Canon City, Colorado. He was one of the giants of all time: 72 feet from nose to tail and 30 tons of flesh and bone. There were other dinosaurs somewhat more gigantic. Moreover, it was reserved for our own mammalian order to generate in the womb of the oceans the largest animal that has ever lived: our contemporary, the 100-foot long, 150-ton blue whale. But Happy, as his local admirers call him, was nearly the largest of land animals. In his massive bones we behold the limits of structure that biological engineering can oppose to the force of gravity that bears us all to earth. His legs are piers; his arched spine, a bridge; his neck, a derrick. The muscles and tendons that articulated these great bones and moved them, along with their own enormous bulk, developed the ultimate tensile strength and contractile power of living fiber.

We can see that Haplocanthosaurus was a placid, if not happy, animal. His neck tapers down to an oddly tiny head that contained a brain no bigger than your fist. That brain—one one-hundred-thousandth of his body weight—could have had little capacity beyond that of a switchboard. Happy was a robot, immune to neurasthenic disease. In fact, a single brain at one end of his 72-foot spinal cord was not enough to handle the traffic of motor and sensory impulses. At the propagation speed of 180 feet per second, it would have taken an impulse almost a full second to make the round trip from stern to stem and back again. To short-cut the transmission time, Happy had a sort of assistant brain, his so-called sacral plexus. Thus:

> *We clearly see by these remains*
> *This creature had two sets of brains;*
> *One in his head (the usual place),*
> *The other at his spinal base.*

In some dinosaurs the sacral plexus was many times larger than the brain. Of these creatures it has been said that they were capable of reasoning not only *a priori* but *a posteriori*. All of which raises the suspicion that the mysterious lower back pain afflicting so many of us today may be, in truth, an atavistic headache.

Necessarily, Haplocanthosaurus and others of his bulk were plant eaters. No other source of food could wait upon their ponderous appetites in sufficient quantity. Yet, considering their huge dimensions, these sauropods were not big eaters. Weight for weight, their metabolism proceeded at a fraction of the rate of ours. Mammals consume most of their food energy in maintaining the heat of their bodies. The elephant thus represents the limit of size attainable by a land mammal, and the poor quivering, chattering shrew has to eat several times its body weight every twenty-four hours. Dinosaurs, on the other hand, could rely upon the world-wide warmth of the Mesozoic age to sustain their body heat. In this sense, they were "cold-blooded," but they were not cold. The enzyme-mediated reactions that summed up to life in their bodies were not very different from our own and required a temperature of about 100 degrees Fahrenheit for ignition. Reptiles in our less favorable times regulate their body temperature by alternately basking and seeking the shade, and they gain and lose body temperature at a rapid rate. A stable internal temperature, according to Edwin H. Colbert, curator of paleontology at the American Museum of Natural History, was one of the great advantages the sauropods attained with size. It took more than a day for the great mass of one of these beasts to radiate away the heat of one degree of body temperature.

With the climate perpetually pleasant, food in abundance, and no capacity for anxiety, it is no wonder that Haplocanthosaurus lived a century or two. Once he had gained something like 30,000 times on his hatching weight few hazards could threaten his survival.

He is displayed here in the presence of one of those hazards: a meat-eating carnosaur. It is a scene that evokes the image of "nature red in tooth and claw." The massive head, the dagger teeth, the prehensile claws, the 30 feet and six tons of poised ferocity, all stir the darkest depths of our unconscious. We must not be misled, however, by the scene and the spectacle we make of it in our imaginations. The sauropod and the carnosaur lived in mutually sustaining symbiosis. Frank Fraser Darling, a student of the community of antelopes and lions on the African prairies, assures us:

Predation is unimportant as a factor regulating the size of animal populations which are, for the most part, self-regulating. . . . The effects [will vary] with the degree to which a population of prey is enjoying optimal conditions in its ecological niche. The complex of predation may be important . . . in conservation of habitat and consequently of prey, by softening zenith and nadir of population oscillations and so lessening the percussive effects on habitat.

Ceratosaur, the carnosaur we see reconstructed here, is a more typical dinosaur. Like man, he is a biped, and there is no doubt that the bipedal habit played a primary part in bringing the dinosaurs to their supremacy in the Age of Reptiles. Haplo-canthosaurus had to drop down onto all fours to hold up his huge bulk. For other dinosaurs the erect posture, counterbal-anced and stabilized by the tail, opened up all the niches of their environment to occupancy. There were dinosaur equiva-lents of the tree-climbing squirrel, the ground-running ostrich, the browsing giraffe, and of the ferret, fox, and lion; for each plant eater had its carnivore.

The most numerous and so, perhaps, the most successful were the dinosaur-antelopes, the duck-billed hadrosaur. His bill is a bony shovel for rooting and scooping in the marshy margins of sea and land that constituted so much of the land area in those warm and watery times when the sea level was so high. It is apparent that hadrosaur ground up a great deal of abrasive dirt along with his food, for his most extraordinary piece of equip-ment is his dentition: four batteries of grinding teeth, packed in the jawbone like cartridges in a rifle clip.

The dinosaurs, numerous and diverse as they were, shared the world with other reptiles. Some, perhaps to make way for the dinosaurs, returned to the oceans: the ichthyosaurs and saurop-terygians became reptilian fishes and whales. Some, the ptero-saurs, took to the sky and became the forerunners of the birds. Others stood their ground and became the forerunners of the modern reptiles, the crocodiles, turtles, lizards, and snakes. In the cover of the underbrush, there lurked also the alert, high-strung, quick-moving protomammal whose descendants were to inherit the earth.

In this catalog of the reptiles, the dinosaurs stand out because they are not among the living. They are extinct. On this account they are often said to be failures. It is easy to expose the fallacy here: extinction is no distinction! Of the 2,500 families of animals that have ever lived, only 800 are represented directly or on collateral lines among the forms alive today. Extinction has served life as a kind of sanitation, clearing the way for the invention of new forms.

It can be argued with more conviction and evidence that the dinosaurs were a huge success. The line flourished for 120 million to 150 million years; some species for 20 million to 30 million years. The great wealth of the dinosaur gene pool is bodied forth in thousands of species, responsive to diversity in the environment and to change as well.

When we consult the dinosaurs for the wisdom that their long experience might shed, we must be careful not to force the answer in the framing of our question. If we would hear these silent bones, we must hold our tongues.

We cannot, for example, find a moral in the story. Until the dawn of human consciousness, the world knew neither good nor evil. The "struggle for existence" was not, as the term implies, the war of all against all. Natural selection proceeded as much by cooperation as by competition. "The survival of the fittest" was secured by branching rather than by pruning. Diversity was self-regenerating; each new life form opened the possibility of life to another. The great extinctions were worked not through pre-emption of the niche by other animals and surely not by predation, but by changes in climate and habitat that challenged the entire existing community of life.

Some such geological cataclysm extinguished the dinosaurs all at once and suddenly—sudden in the geologic sense of spans of millennia or centuries—at the end of the Age of Reptiles. As for the dinosaurs, they probably did not notice that anything was going wrong, and the last one was not even lonely.

The Age of Reptiles stands as a pivotal, central chapter in the history of life. The passage of these lords of the earth ought to persuade us that change is inevitable. Arising from the clay in all their triumphant vitality, testing the very limits of tissue, and then vanishing in all but human memory, the dinosaurs testify that life is an ongoing historical process.

In their place in history, midway on the scale of capacity and complexity but recent in time, they show us also that change accelerates. Each major period in the history of life is of about half the duration of the preceding phase: once the rotation of the earth in the sunlight had brought the synthesis and assembly of single-celled organisms, it took about 3 billion years for these to give rise to multicelled forms; a half-billion years for these forms to evolve into chordates; 300 million years to fashion the vertebrates and begin the conquest of the land; 150 million years for the Age of Reptiles to run its course; and 70 million years for the Age of Mammals.

Now the Age of Mammals has come suddenly to end in the swiftest and grandest cataclysm of all time; within the last 10,000 years: the ascendance of man. The Age of Man—if it is to have half the duration of the Age of Mammals, if we are to hold our heads up to the dinosaurs—should run 35 million years.

While the record shows that change is inevitable, it also shows, in Julian Huxley's carefully hopeful words, that progress is possible. This is, in fact, a declaration that man must make. As the ultimate expression of the progress of life, he is the only living thing qualified to make it. He can see that progress has objective definition in the record: progress is increasing capacity for self-regulation and corresponding independence of the environment. Such a major step as the conquest of the land, for example, required the liberation of the reproductive process from dependence upon the ocean; the water is self-contained in the egg. Similarly, the mammalian self-regulation of body heat frees the mammal from dependence upon environmental temperature.

Such liberation of life forms involves a nice balance between the efficiency attained by specialization and the potentiality maintained by less specialized forms. The dinosaurs, each so comfortably if not elegantly adapted to its niche, were caught in dead-end dependence upon habitats that vanished. It was an unspecialized reptile that became a mammal and survived. In the succession that followed, it was an unspecialized mammal that became a primate, not an ungulate. It was an unspecialized primate that became not a gorilla but man.

That primate was the first toolmaker. He lived nearly 2 million years ago in the Rift country of Africa, where fragments of the bones of his hands have been found along with the stone tools that they made.

Toolmaking is necessarily purposive behavior. Some feedback loop had to close in the circuits of the cerebral cortex, integrating the motor, sensory, and memory faculties of the brain. The toolmaker, by definition and by physiology, was a future-dweller. He became, therefore, a truth-seeker; what he learned now could secure his safety later. He became, by the same token, an inventor of value; things were either good or evil to purpose. It is true that the record as to the bones of hands and skulls is scanty. But the record is rich in fossil behavior preserved in the stone tools. Long before *Homo sapiens* put in his appearance, the record shows, his forerunners were conjuring with truth and value.

In the emergence of our species, evolution has transcended itself. From independence of the environment man has gone on to seize control of it. Never in history has any lifeform come so swiftly into domination of the earth; never before has a single species dominated all others. It is apparent—especially since the existence of man places all other forms of life on probation—that man is the terminus of evolution. No successor is capable of evolving from some other form to consciousness again. Evolution, from now on, is a moral enterprise, subject to man's conscious purpose and direction.

To all of this it may be added that the only mode of extinction that threatens man's tenure for geological ages to come is self-extinction.

Now, the scandalous fact is that this new lord of the earth is actually contemplating self-extinction. Our predecessors and our contemporaries displayed here in this Hall of Nature, for our pleasure and instruction, must look upon us with bewilderment and horror.

Why, even the brainless dinosaur would have sense enough to say:

Cherish life. You have barely begun; you have all the time in the world. No value or purpose you have yet framed can justify your terminating the possibilities of life you have just now taken into your hands.

II Common Sense

The Heritage of
Science in a Civilization
of Machines

THE Apollo moon voyages have served to popularize to a world-wide audience the basic propositions of celestial mechanics that launched the scientific revolution nearly four centuries ago. As one employed in the popularization of science, I must count this as a gain. At the same time, I fear, these extravagant adventures have deepened the misunderstanding of science that prevails in our time. The revolution still has a long way to go.

The voyages are, above all, a triumph of "systems engineering," the orchestration by advanced computer and communications technology of the far-flung land, sea, air, and space crews engaged. To uncounted numbers of onlookers the demonstration confirms the vision of science as a group-think enterprise in which individual human personality has no useful role. The shoe box or two full of moon rocks fits equally the picture of the scientist as fact collector, with none of the artist's creative engagement in his subject. For the out-of-scale relation between the effort and the yield, the explanation must be that science is truly value-free.

Such misconception is the concern of the four essays in this section. They will never reach Apollo's wide audience. But they were addressed to the source of the schizophrenia that permits the nonrational choice of national purpose. This is the peace-making agreement in the faculties of the arts and sciences to the effect that there are two ways of knowing. The corresponding division of labor relieves the scientist of uncomfortable moral responsibilities and delegates those cares to the scientifically illiterate.

Something of the spirt of the academic compromise was implied in the title of the symposium on "The Heritage of Mind in a Civilization of Machines," held in October, 1962, to celebrate the 150th anniversary of Colby College, at Waterville, Maine, to which I contributed the first of these essays. The reader will find here, incidentally, that events in the interim have lighted up a cloud in my crystal ball. I had accepted the pessimists' judgment of the moon-orbit strategy then being chosen for the Apollo mission. It turns out, as only the proponents of that strategy believed in 1962, that the subtle tasks of navigation and control proved more amenable to engineering than the brute-force problems of propulsion.

I<small>N A</small> civilization of machines, what is the heritage of science—other than the machine itself?

The question is plainly rhetorical. It is the preliminary question that clears the approach to higher ground. It sets aside our contemporary obsession with ways and means and opens the path to exploration of the realm of ends and value. The door behind is shut with the statement that the advance of science has somehow outrun the progress of philosophy and so the time has come to rediscover the sources from which men once drew motivation and purpose.

Now, I have unwound the rhetoric in this fashion because I want to show you that the question has another answer: The heritage of science encompasses value as well as machines. What is more: Science offers the only source of value open to twentieth-century man. For there can be no retreat from the irreversible advance of human understanding. The rediscovery of the past can lead us back only to the present. It is here that we must make our stand.

In the recent history of machine civilization, I will admit, there has scarcely been a less auspicious moment for such an affirmation. The succession of Ages that celebrates the passage of modern times has ushered us suddenly, without warning, from the Age of the Atom into the Age of Space. If the Atom confounded us with the schizophrenia of hope and despair, then Space has made us look into the void within as we prepare to venture into the harshly lighted vacuum beyond our sky. To what end, to what other ill-considered purpose will we be led by the next of the Great Ages of the civilization of machines?

If we are to seek enlightenment in science, as I urge, then let me begin with a caution. One who would cherish the heritage of science in the space age must learn to distinguish between science and science fiction. Here on Mayflower Hill, you may not have had the opportunity to make firsthand acquaintance with the marvels of this age. Those of us who have come here from the world outside, especially those who have had recent occasion to visit Florida or Southern California, can testify to you that rumor has not improved upon reality. The gigantic rockets, the awesome discharge of violence that propels them, and the intricate systems that navigate them—all make fact of fiction. To this I must add: the people of the space age are no less improbable than their machines.

By way of illustration, let me reconstruct for you an event, symbolic of this period, reported in the *New York Times* just two weeks ago: the President of the United States has arrived at the George Catlett Marshall Space Flight Center at Huntsville, Alabama, in the course of a photographic visitation to the nation's space facilities. He is ushered into the office of Dr. Wernher von Braun. There, the *Times* reported:

As Dr. von Braun, director of the center, was showing Mr. Kennedy a small model of the advanced Saturn C-5 rocket, which will be used for the lunar shot, he turned to the President and said:

"This is the vehicle which is designed to fulfill your promise to put a man on the moon by the end of this decade."

He hesitated a second, then said: "By God, we'll do it!"

Now the real-life President, who has committed some $30 billion of other people's money to this lunar enterprise, may have taken some reassurance from the manly declamation of his space captain. But, I imagine, he must have found greater reassurance during the course of that day in the sight of an actual Mercury Atlas rocket, poised on its launching pad in preparation for the six-orbit flight of Commander Walter M. Schirra, and in the breathless drama of a countdown for the next day's launching of a Titan II, the booster that is to place two astronauts on orbit in the womb of the Gemini space capsule. And the President probably felt that prospects were even brighter when he was brought into the presence of a full-scale mock-up of the F-1 rocket engine, five of which clustered in a single booster will start the lunar explorers on their voyage, hopefully to return.

For the President and the electorate, however, this day of promise and anticipation was marred by a curious episode that took place a little later as the presidential party made its way through the Saturn assembly plant. According to the *Times*, the procession was brought to a halt and:

President Kennedy stood by, obviously somewhat nettled, as Dr. Jerome B. Wiesner, his principal science adviser, engaged in an argument with Dr. Wernher von Braun, director of the Marshall Space Flight Center, over the best way to reach the moon.

Dr. Wiesner, according to those who overheard the discussion, disagreed with the method favored by the National Aeronautics and Space Administration and Dr. von Braun. Under this plan, a lunar "bug" with two men in it would be detached from the parent capsule while in orbit around the moon, and would descend to the lunar surface.

One man would remain in the parent capsule. Later, the bug would take off from the moon and rendezvous with the circling parent capsule. The bug would be abandoned before the capsule returned to earth.

Dr. Wiesner is understood to favor a method that was earlier agreed on and then discarded. Under this method, a lunar rocket would be formed by a rendezvous of parts orbiting the earth. It would blast off from the parking orbit and land on the moon. Later it would take off and return to earth again.

Now the practiced reader of space fiction will see immediately the logic of this episode. It supplies the element of human conflict that is so necessary to lend drama to the otherwise inexorable performance of machines. On the other hand, the President's discomfiture is also deserving of sympathetic consideration. Men of affairs are often exposed to anxiety by the ambivalence and uncertainty of the men whose concern is with things. The engineer or the scientist—the thing-man—always tries to hedge and qualify what must be in the end a clean-cut decision. He has second thoughts; he squeezes design changes into the last minute. This kind of behavior costs money, and, worse yet, it stalls the forward march of great enterprises. Somehow technical experts fail to comprehend the larger context. On occasions such as this, when they are so heedless as to conduct their disputations in the presence of reporters, their preoccupation with their narrow technical concerns can be downright embarrassing.

The rest of us must nonetheless be grateful for this episode. Here, in a casual argument between two of the principal science-policy makers of our government, we gain a giddy insight into the virtuosity with which man may now summon and manipulate the forces of nature. The episode gives us still another insight: into the "closed politics," as C. P. Snow has called it, by which an increasing number of decisions are taken in our increasingly complex world. Decisions issuing from contests of brain and will conducted in private behind closed doors and frequently under the cloak of military secrecy deploy the human and material resources of the entire nation. They may one day commit our lives as well as our fortunes, without prior notice or consultation. Yet the advocate of open politics is hard put to show how these decisions might be made otherwise. So very few of us are qualified by talent or training to contribute to the deliberations from which these decisions come.

What is more, the brief history of the Space Age would suggest that something has gone wrong with the traditional decision-making machinery of our government. The attempt to put a man on the moon by the end of this decade was undertaken in broad daylight. It is true that the proposal originated in secret, as an imagination-capturing venture of a young new administration. The administration had to enlist the approval of the Congress and of the electorate, however, by something approaching the open procedures prescribed by our Constitution. Yet the sense of this decision to go off to the moon remains as obscure as the argument between Wiesner and Von Braun about how to get there.

The motives that ruled the decision to proceed with the project had little relevance to its inherent values. Project Apollo was undertaken, before all else, as a ploy in the cold war—the Russians may yet beat us to the moon. It is thus a paramilitary enterprise; by brandishing their missiles out in space the two Great Powers seek to enhance the credibility of their deterrents. Next, one must mention the economic compulsion; this gains ascending importance as military expenditures yield a diminishing stimulus to the economy. The last and least of the motives was the increase in human knowledge that a successful round trip may yield.

It is impossible, of course, to put a price on the things we may learn from a visit to the moon. I, for one, will argue that the scientific motive may be taken as sufficient. Disturbing questions must nonetheless be asked. It is, for example, a moral as well as a technical and financial question whether we should send men instead of robots to the moon. We must also ask: Why in a decade? Finally, we ought to examine the order of priority which we have assigned to the pressing demands on the capacities and resources of our country.

These are questions on which the public can and ought to be consulted and to which public discussion may bring wisdom and humanity. The same may be said even of the highly technical issue that divided Wiesner and Von Braun. One of the two routes to the moon looks more like a round trip for the astronauts. The other promises an earlier launching. It is easy to guess which route has been chosen by the method of closed politics, and it is likely that open politics would choose the other.

Citizenship in the machine civilization, therefore, calls for wary discount of the marvels of science fiction. Some people think they have a clear-cut test; the space feats of the Soviet Union, they say, are fiction. The Russians simply faked the "beeps" alleged to have come from the first Sputnik, forged those photographs of the far side of the moon, and generated the TV picture of Nikolayev lounging in his space cabin by straightforward electronics. A closed society is, of course, vulnerable to such charges. The logic of this line of argument, however, requires us to believe that the Russians are incompetent in space and at the same time incomparable masters of the techniques of mass communication. Carried just a step further, it throws our own space feats in doubt. How do we know the entire space pageant is not a gigantic hoax—an optical illusion created by a conspiracy of the press, the radio, and television through which the rest of us have been sharing vicariously, at third hand, in this latest adventure of mankind?

How, in this age of fact made fiction, do we know what we think we know? Most of us are ready enough with our knowledge, especially those of us who have had the privilege of higher education. Ask your well-informed contemporary, and he will freely render an acceptable account of modern cosmology: our earth is an undistinguished (except for our presence on it) planet of an undistinguished star located in an undistinguished region of an undistinguished galaxy in—he will not fail to add—an expanding universe.

Ask him, then, how he comes to know all this. He will doubtless lead off with the observation that the rising and the setting of the sun shows our earth to be turning on its axis as it travels on its orbit around the sun. His assurance on these points will be dampened, however, when you remind him that men as intelligent as ourselves have watched the sun rise and set for more than 25,000 years and, until very recently, held that the sun revolved around the earth. In the end, it will turn out, he grew up knowing these things. He learned them by parental instruction, in school, and from books; he knows what he knows because someone told him it was so. His knowledge, in sum, rests upon the same foundations of authority and faith as that of ancient and primitive man.

In some ways the typical cosmology of the ancient or the primitive man is more honorable and simplistic. As to that portion of the universe which he beholds with his own eyes, he makes no complicated correction of what he sees. The earth is flat and stationary and it is roofed over by the revolving dome of the firmament. This was the cosmos of the Bible and of Homer. At different times and places it has been variously embellished by legends taken on faith. So long as men were bound to the soil, the sun played the central life-giving role in the mythologies that illuminated the all-encompassing mystery. The moon, by coincidence of the menstrual and the lunar cycles, was commonly assigned jurisdiction over fertility. In the cosmology of ancient Greece, the universe outside the celestial dome was filled with fire that shone through the tiny windows of the stars. In the cosmology of the medieval fathers, fire was infernal, not celestial, and it roared in the bowels of the earth; the still deeper depths of hell were locked in eternal ice. From one age to the next, the masses of men were acquainted with these marvels through the offices of their prophets, priests, and philosophers. These learned and inspired teachers came by their knowledge in turn through vision or revelation or other ecstasy of the imagination.

The question of how we know what we know today is rightly framed in the context of cosmology. Astronomy is historically the first of the sciences because the firmament was the realm of nature first susceptible of precise and repeated observation and measurement. The straight lines described by rays of light made possible the exact measurement of angles; the repetition, from day to day and year to year, of the sidereal cycle made it possible for different men at different times and places to make the same measurements and to compare their results.

Early in history, one can discern the divergence of science from other methods of philosophy. Observation and measurement engage the imagination in external reality. The results achieved by this method make a difference. Each observation, as it is made and verified, leaves its mark indelibly upon the world. Step by step, nature is transformed before the eyes of man. With measurement there comes prediction and, with prediction, control. The method is accumulative, and, over time, the rate of accumulation speeds up; this is because each addition to knowledge proposes new questions that no one would have thought to ask before.

As early as the sixth century B.C., for example, it had been observed that the wheeling of the sky turned the constellation of the Great Bear through a complete circle around the Pole Star when observed from the latitude of Greece and carried it below the horizon when observed from Egypt. To Anaximander this observation suggested an ingenious amendment of the Homeric cosmology: the surface of the earth was curved in the north-south direction; that is, the flat earth became a section of a cylinder with its axis pointing east and west. The Pythagoreans, in their commitment to aesthetic values, warped the cylinder into the more "perfect" shape of a sphere. But they also had another set of observations to account for—the shadows cast by the earth upon the moon—and they had sufficient sophistication in geometry to interpret what they saw. In the third century B.C., with the sphericity of the earth established, Eratosthenes was able to undertake the measurement of its size. He observed that a stick of a measured length stuck upright in the sand at Syene, on the upper Nile, cast no shadow at noon on the day of the summer solstice. At the same hour on this day in Alexandria a stick of the same length cast a shadow of a length that showed the sun to be at an angle of $7° 12'$ from the vertical at that latitude. From this observation and calculation, plus a sufficiently exact measurement of the distance from Syene to Alexandria supplied by professional runners, Eratosthenes was able to measure the diameter of the earth to an error of only one part in one thousand—within 70 miles of the modern equatorial diameter of 7,927 miles.

Measurements of this kind supply the decisive test of knowledge in modern times. They distinguish what we know from the knowledge carried in the heads of our forebears. Most of the knowledge of most of us still comes to us by hearsay. We can take confidence in this kind of knowledge, however, because we know that it rests upon observation and measurement made by experiments that have been repeated more than once; experiments that we ourselves could perform, had we the time, the tools, the wit, and the inclination; experiments that we can perform, in any case, in our heads. When we accept the authority of a scientist we do so in the assurance that his peers will expose him if he ventures into fraud.

Though the experiment is thus decisive to knowing, it is the imagination that drives the expansion of knowledge. The act of the imagination that explains the latest experiment and proposes the next one is the creative act in science. By such an act, Copernicus placed the sun at the center of the universe. The imaginative quality of this bold stroke is revealed as much in the errors that it contained as in the success it won. Against the background of the fixed stars, the planets trace strange, looping courses, now advancing, then retreating and then again swinging westward toward the setting sun. Moved by aesthetic considerations to picture the planetary orbits as circles, Copernicus found it necessary to distort his model of the solar system. He made the sun the center of the earth's orbit, but he had to postulate another point, K, as the center of the circular orbits of the five other planets, in order to account for their peculiar apparent motion in the sky.

The physics for this new cosmology was supplied by Galileo's concept of inertia, or gravity, as inertia came to be called later in its universal Galilean sense. His *Dialogue on a New Science* describes how he undertook to measure the acceleration of gravity acting at the earth's surface on a falling body. He rolled a ball down an inclined plane and timed its passage from interval to interval as it rolled. By thus diluting "inertia," as he put it, he brought the acceleration of the ball within the reach of his instruments for measurement of time. From a critical reading of his account of this experiment, however, historians of science today have developed the strong suspicion that Galileo never really performed it. His results are so perfect that it must have been a "thought experiment," performed in his imagination.

In the great world system with which Newton crowned the first century and half of modern astronomy, the imagination bridged the void between the few isolated corners of the universe that men had measured and brought them into orderly relationship with one another and with regions still unexplored. In a few brief statements of incandescent clarity and rigor, Newton laid out the lifework of twelve generations of science.

The scientists who extended the Newtonian equations and the engineers who employ them today take assurance from their beauty as well as from their demonstrated power. The Greek aesthetic prevails in science. Hypotheses and theories must have the elegance of simplicity and the symmetry that brings equations into balance. In the end, however, the statements of the scientist must stand the test of verification by repetition of the observation or experiment and by measurement. As to what such a test may show—even though all concerned are able to agree upon the conclusion—the scientist can trust no authority but his own judgment. This is the first value taught by science, the touchstone of all value: the autonomy of the individual.

The ultimate contemporary proof of the power of this method of philosophy is, I suppose, our arrival in the Space Age. One may say, at least, that the ideas of Copernicus, Galileo, Newton, and their successors have now been popularized on a scale appropriate to their grandeur. Men have repeated the celestial experiment that placed the moon on orbit and have adorned the sun itself with new satellites.

The triumphs of space technology involve, of course, much else in addition to celestial mechanics. The Promethean theft of Galileo and Newton placed the universal laws of motion at man's command on earth. Since then, physics has triumphantly invaded every realm of natural philosophy. The daring ventures into the depths of space now in preparation invoke deep and diverse knowledge of the nature of matter and the forces of nature, of life processes, and even of human perception and behavior that no single human mind as yet has comprehended.

A true comprehension of the heritage of science would show that much else is possible. It is the possible, however, that confounds the questions of ends and value. From age to age in the past, it was always the newly possible that upset the settled arrangements of human society and controverted the prevailing justification of God's ways to man. If this means that value itself is subject to change, it does not follow that value is merely relative—that pushpin is as good as poetry. The good and the just are not, on the other hand, absolutes awaiting more perfect revelation. They are human inventions; they are the first derivative of truth-seeking and toolmaking. At this juncture in history, it is well to observe that the protests of ennui and surfeit come exclusively from the world's well-fed. There are no such dilemmas about value in the bellies of the hungry.

As we compare our estate with theirs—whether those presently deprived or the overwhelming numbers of our predecessors who lived and died in destitution—we should also recall the circumstances in which the values of our culture were first asserted. Self-government and science are coeval because both are the enterprises of self-responsible and independent men. Liberty became the right of all members of society and not a privilege of class only when the progress of machine technology displaced human muscle as the prime mover of economic life. Equality only now becomes a social reality as the abundance generated by automatic production subverts our economic institutions. Excellence, the perfectibility of the individual, is a goal to which increasing numbers of our younger generation may aspire—today, measured by enrollment in our colleges, something more than twenty percent.

The example of America has, more than any other single factor, set the hungry nations of the world on the course of industrial revolution. If, in the course of their inevitable, concomitant social revolutions, they do not seem to share our regard for the values of democracy, that is because food, clothing, shelter, and the technology that produces them must come first—and did come first in the history of our own social evolution, as that history will show when it is honestly written. The difference between today and yesterday is that the technology to meet these material demands—and so to minimize the human costs of the next industrial revolutions—is at hand. Among the pressing demands upon our unexampled capacity for the generation of abundance I would give high priority to aid for the developing countries. Others have computed the number of universities, of steel mills, dams, and acres of irrigated land that could be created by applying a different order of priority to the treasure and talent now lavished upon space ventures—and on the warmaking capacity to which they are incidental.

By this tragically roundabout reasoning, if by no other, we should at last be able to persuade ourselves that we can make this earth a fit habitation for our species. We must come to this undertaking soon. Our increasing numbers, the wasting of the earth's treasures, and the overhanging threat of weapons of mass destruction compel us to make peace with one another and with our natural environment. It is time we embraced the heritage of science in this civilization of machines. There are already eight barren planets on orbit around the sun.

The Common Sense of Science and the Humanities

WHILE Harvard College holds the center of the stage at commencement every year, the university does not slight its other schools and their alumni. This essay was contributed to the festivities organized for the alumni of the Graduate School of Arts and Sciences in June, 1967, on the day before the most resplendent annual rite of our democracy.

THE sun that presides without fail at every Harvard commencement starts up each year another round of a sedentary yet elevating outdoor sport. All that a player needs is a copy of the commencement program and a shady seat in the Tercentenary Theatre. No training is required; for the players are all spectators. The name of the game is the decoding of the titles of theses.

Bachelors who qualify for honors and Masters all write theses. But the honor of listing by title goes only to theses that earn doctorates. This discrimination on the part of the authorities makes the game playable within the few solemn moments of the commencement ceremonies proper. The task of the spectator-sportsman is to tell what each thesis is about.

Now, the 1967 titles are at the printer's. They will not be placed in action until tomorrow morning, when the Honorable and Reverend Board of Overseers shall have voted their consent. We can get in a little practice here, however, with the 1966 commencement program in hand.

Some titles are transparently self-explanatory: "Count Dmitri Tolstoii and the Russian Ministry of Education 1866–1880"; "On the Distribution of Galaxies in Space"; "Wish Fulfillment and Psychic Energy in Psychoanalytic Theory"; "Reactions of Polynuclear Hydrocarbons with Pyrimidines"; "Tariffs and Real National Income: A Policy Model."

An occasional title invites reading of the thesis, footnotes and all: "Downstairs from the Upstairs: A Study of the Servants' Hall in the Victorian Novel."

Some titles are all too explicit: "Presentation of a Referendum to the Citizens of Mount Vernon, Illinois, to Increase the Tax Rate Levy Limit on the Education Fund from Eight Cents per One Hundred Dollars Assessed Valuation to One Dollar and One Cent per One Hundred Dollars Assessed Valuation."

Many benefactors of the university would find a ready interest in: "Corporate Repurchases of Already Outstanding Common Stock."

But the same good people might fail to recognize their stake in: "Strategic Situation Theory: A Bayesian Approach to an Individual Player's Choice of Strategies in Noncooperative Games."

All who hold the Bachelor's degree from this institution, men and women respectively, will be attracted by: "The Achievement Motive and Differences in Life Style Among Harvard Freshmen," and "The Career Within a Career: The Freshman Year at Radcliffe."

But what do you make of: "*Lay la Freine* and the Breton Lai"? "Repressive Defensiveness, Identity and Cognitive Style"? "Random Methods for Nonconvex Programming"? "Archplot as

Sphragis in the Plays of Euripides"? "Toward 'Uhuru' in Su-
kumaland"?

In what field of learning would you confer the doctorate on:
"Conformal Invariants for Condensers"? "Wessel Gansfort and
the Art of Meditation"? "Magnetic Relaxation in Rare Earth
Doped Garnet"? "The Number of Generalized Runs in a
Markov Chain Sequence of a Fixed Length"? "Where Terms
Begin: Giambattista Vico and the Natural Law"? "An Empiri-
cal Comparison of Magnitude and Category Scaling Procedures
Applied to Nonphysical Stimuli"? "Guevara Across the Cen-
turies"?

As the eye scans the list, questions of a deeper kind come
forward. From which of these new Doctors will we hear again?
Which of these titles mark the departure of original and enterprising scholars and scientists? In that Class of 1966 was there
another William James, a George Lyman Kittredge, a Roger
Bigelow Merriman, a Percy Bridgman, a Howard Mumford
Jones?

If we cannot answer, it is not alone because a title, however
explicit, must fail to convey the breadth and depth of the thesis.
Nor must we tax ourselves with our ignorance. The expansion
of knowledge implies not merely more of the same but increasing diversity of knowledge. If the expansion now proceeds as an
explosion, we can count it a more hopeful one than others we
have lived to witness in these explosive times.

We may find reassurance as well in contemplating the unity
that binds the diversity of human enterprise we behold here.
These young scholars have set off on diverging pathways in
pursuit of the same high purpose. As Charles William Eliot
declared at his inauguration, "This university recognizes no
antagonism between literature and science, and consents to no
such narrow alternatives as mathematics or classics. . . . We
would have them all, and at their best." The passage of a century has shown that the expansion of the universe of learning
may be contained by concurrent increase in the gravitational
attraction of the university at its center.

At this point, however, I am compelled to admit that literature and science, mathematics and classics, went more comfort-

ably together in Eliot's time. The historian George Ticknor rejoiced in the great things the naturalist Louis Agassiz was doing for Harvard. "By making Natural Science move," he said, Agassiz made languages, history, and literature follow. "Natural Science has tended to open Harvard College; to make it a free university, accessible to all, whether they desire to receive instruction in one branch or in many." The "and" in Faculty of Arts and Sciences was in those days a simple conjunctive.

Today, I fear, that little word has come to stand as a disjunctive. It seems to imply Arts "as distinguished from" Sciences. There is abroad in the world the notion that men have two ways of knowing.

Two kinds of language speak for two realms of distinct, if not opposed, concern. The vocabulary of the one claims such terms as synthesis, grace, quality, intuition, feeling, beauty, and mind; while the other speaks in terms of analysis, precision, number, logic, fact, truth, and brain. The first evokes enduring concerns; the second sounds like hard work. In other words, a university has two faculties: one is occupied with ends, the other with means.

This is at least the popular view, as may be judged by the words of the nation's foremost practicing moral philosopher. When Lyndon B. Johnson put his presidential signature to the legislation establishing the National Foundation for the Arts and Humanities, he declared: "We need science to make our goods. We need the humanities to give us our values."

To the extent that this notion prevails in centers of power— in government, in the great foundations, and in our universities—it is as pernicious as it is mistaken. It is probably not a new idea.

As if to counter such a notion, Eliot was moved to declare in that same inaugural address: "A university is not closely concerned with the application of knowledge, until its general education branches into professional. Poetry, philosophy, and science do indeed conspire to promote the material welfare of mankind; but science no more than poetry finds its best warrant in its utility."

Evidently, the popular understanding of science still stands in need of clarification. James Bryant Conant, during the last years of his presidency, addressed himself actively to "the question of how we can in our colleges give a better understanding of science to those who have no intention of being scientists. . . ." He rightly judged that "the remedy does not lie in greater dissemination of scientific information among non-scientists. Being well informed about science is not the same thing as understanding science. . . ."

In the general education courses that he designed and taught, Conant undertook to demonstrate what he was careful not to call "the scientific method." Out of his national service in World War II, Conant drew a martial metaphor; he declared that what needs understanding is "the tactics and strategy of science." But he took care to show his students and his readers that the methods (plural!) of science constitute nothing more mysterious than the painstaking and rigorous extension of common sense.

There is, of course, only one way of knowing anything. That is: common sense.

The dust has long since settled on the question of the grounds of knowledge. If the conclusion of the ancient and lofty debates in epistemology has a deflating, Yankee sort of quality, that is perhaps because Harvard is one of the places where it was thrashed out. Along with the other Cambridge and with Vienna, this university is one of the principal springs of the mainstream of modern philosophy. The Harvard figures in this movement are no less central than Karl Pearson, G. E. Moore and Ludwig Wittgenstein, Ernst Mach and Rudolf Carnap.

As early as 1878, Charles Sanders Peirce staked out the claim for common sense. Peirce was a product of the Agassiz revolution at Harvard and the son of Benjamin Peirce, professor of mathematics and founder of the Harvard Observatory. He quit academic life to work as a practicing geologist in the U.S. Geological Survey. Writing in the *Popular Science Monthly*, then published by the American Association for the Advancement of Science, Peirce showed his readers "How to Make Our Ideas Clear."

"There is no distinction of meaning," he said, "so fine as to consist in anything but a possible difference of practice. . . . It appears, then, that the rule for attaining the [highest] grade of clearness of apprehension is as follows: Consider what effects, which might conceivably have practical bearings, we conceive the object of our conception to have. Then, our conception of these effects is the whole of our conception of the object."

Peirce called this sensible piece of advice "pragmatism." Under that name its implications were explored and organized into a system of philosophy by the poetic genius of William James.

Against the metaphysics that "always hankered after unlawful magic," James showed how the pragmatic approach could "bring out of each word its practical cash-value." In place of the absolutes that close the metaphysical quest, the pragmatic "attitude of looking away from first things, principles, 'categories,' supposed necessities; and of looking towards last things, fruits, consequences, facts" sets up "a program for more work, and more particularly . . . an indication of the ways in which existing realities may be changed."

Against the "hurdy-gurdy monotony" of the "fearful array of insufficiencies" propounded by dogmatic materialists of the type of Herbert Spencer, James pressed the same critique. "The truth of an idea is not a stagnant property inherent in it. Truth happens to an idea. It becomes true, is made true by events. Its verity is, in fact, an event, a process: the process namely of its verifying itself, its verification."

In the 1920's and 1930's, pragmatism received a rigorous restatement in the "operational analysis" of Percy W. Bridgman. One of the last solo practitioners in physics, in the days before physics became "Big Science," Bridgman won the Nobel prize for his lonely work in the behavior of matter under high pressure. He is remembered also as the stoic of the philosophy of science.

Summarily stated, the essence of the Bridgman attitude is "that the meanings of one's terms are to be found by an analysis of the operations which one performs in applying the term in concrete situations or in verifying the truth of statements or in finding the answers to questions." The rule holds for the "non-physical . . . operations of mathematics or logic" as well as for the conventional operations of the laboratory.

In his wintry style, Bridgman discounted the suggestion he had propounded "anything new or definite enough to be dignified by being called any kind of 'ism.' " Operational analysis, he declared, "is more of the nature of an art, to be learned by the practice of it and by observation of the practice of others."

Where men have sought the Absolute in the order of nature, operational analysis finds Uncertainty. The terms "velocity" and "position" ascribed to a particle do not denote properties inherent in the particle but experiments to be performed upon it.

The *modus operandi* by which the uncertainty gets into the situation is through the act of observation—the electron cannot be observed without bouncing an atom of radiation from it or doing something equivalent; and, whatever the process of observation, the motion is interfered with. The essential fact is not that the act of observation interferes with the motion; if this were the only effect we could allow for the amount of interference by calculation. The essential fact is that the act of observation interferes with the motion by an unpredictable and incalculable amount.

This, in Bridgman's succinct prose, is the Heisenberg Uncertainty Principle. From the principle, it follows "that in the realm of small things the law of cause and effect does not operate."

Bridgman recognized that this conclusion "is likely to awaken in many persons the most active hostility." But the reason is not "that the physicist is either lazy or a quitter." The essential fact is "that it appears to be due to a law of nature."

By way of commiseration Bridgman remarked: "Hundreds of years of attempting to find inside our own heads the necessary pattern of the external world has proved a dismal failure; we have come to feel that we can only take experience as it comes and must try to get our thinking into conformity with it."

Science, in this characterization and in Bridgman's own terms, is a more "private" than "public" activity. It is a long way from the solemn authority of the impersonal third person and the passive voice in which the work of science is formally reported. As in any other creative activity, the central actor in science is "the simple perpendicular I."

"The process I want to call scientific," Bridgman wrote, "is a process that involves the continual apprehension of meaning, the constant appraisal of significance, accompanied by the running act of checking to be sure that I am doing what I want to do, and of judging correctness or incorrectness. This checking and judging and accepting, that together constitute understanding, are done by me and can be done for me by no one else. . . . They are as private as my toothache, and without them science is dead."

Poets, novelists, composers, painters, and sculptors will all acknowledge this experience as their own. We owe to the astringent intelligence of Richard von Mises a clear exposition of what it is that is common to the aims and methods of the arts and sciences and that brings artists and scientists to share the same pangs at crucial hours in their work. Professor of aerodynamics and applied mathematics here from 1940 to 1953, Von Mises linked Harvard pragmatism and Viennese positivism in the mainstream of philosophy. "Connectibility" was his watchword, and he gave it spacious meaning. "The aim of the intellectual endeavor of man may in the last analysis," he said, "consist in the attempt to arrive, for all phenomena that are of some interest, at a description that is connectible across the boundaries of all fields."

Against the view of certain positivists in the Vienna Circle as well as the chorus of the "negativists," he argued that science and art are more closely related than either set of adversaries would have us believe. Capacity for intuition, vision, feeling— so widely claimed for the artist—plays its role from moment to moment in the work of the scientist. He proceeds from similarities and analogies, apprehended by the same faculties of eye, hand, mind, and heart, to the framing of statements about the subject at hand that can be verified by logical proof, by observation, or by experiment performed by himself or by others. Conversely, Von Mises observed: "Every work of art can be considered as a theory of a specific small section of real life—at any rate, more significantly as a theory than as an imitation or reproduction. . . . A work of art is neither true nor false [that is: not verifiable], but it can, to a greater or lesser degree, agree with our experiences, previous or future." In another context, he declared: "Many works of poetry really make one confident that, at least in certain areas of feeling, there is hardly any experience that is not open to linguistic expression."

The main task of philosophy, according to Von Mises and his colleagues in the Vienna Circle, is "improvement of the use of language." Language is invested with connotation, unspoken meaning, and downright magic; "the 'logic' stored in our language represents a primitive stage of science." Thus the metaphysician thinks "that a word, e.g.: the word 'justice,' corresponds independently of all conventions, to some specific entity, and he seeks to 'discover' this entity. . . ." It is no wonder "that at a very early time the ideal of an unchangeable 'external' truth made itself felt."

Men have conjured for a long time with a class of assertions that seems to satisfy this ideal: the theorems of mathematics. To put the matter in an unfairly simple-minded way, it may be said that the inexorable outcome of the addition of two and two laid the foundations for the resurrection by Emmanuel Kant of the claim for *a priori* knowledge of the absolute. The original positivists, John Locke and David Hume, had already shown these ambitions are illusory. But Kant's advocacy, fortified by his undoubted powers as a mathematician and theoretical physicist, kept the issue open for more than another century.

The issue was settled only when Wittgenstein, building on the work of Bertrand Russell and Alfred North Whitehead, showed that the theorems of pure mathematics and of logic have nothing whatever to say about the world of experience. They are empty tautologies. The certainty of these statements, in other words, derives from the fact that they merely restate their premises in their conclusions.

Willard V. Quine now occupies the chair of common sense at Harvard—more formally, the Edgar Pierce professorship of philosophy. Of his assignment Quine says: "Philosophy is in large part concerned with the theoretical, non-genetic underpinnings of scientific theory; with what science could get along with, could be reconstructed by means of, as distinct from what science has traditionally made use of." He could not have chosen a more trying time to undertake this assignment. For science is just now learning to contend with an uncertainty uncovered in the operations of logic no less disconcerting than the uncertainty found in the processes of nature.

In 1952, Harvard bestowed the degree of Doctor of Science upon the German mathematician Kurt Gödel. Mr. Conant's citation of Gödel read: "Discoverer of the most significant mathematical truth of this century, incomprehensible to laymen, revolutionary for philosophers and logicians."

To Quine we are indebted for the explanation: "What Kurt Gödel proved . . . was that no deductive system, with axioms however arbitrary, is capable of embracing among its theorems all the truths of the elementary arithmetic of positive integers unless it discredits itself by letting slip some falsehoods too. . . . Every such system is therefore either incomplete, in that it misses a relevant truth, or else bankrupt, in that it proves a falsehood."

The dust has not yet settled on this catastrophe. The Gödel proof does not imply for one instant, however, that logic and mathematics lose their role in the structuring of the thinking process. On the contrary, any deductive system may be cured of its bankruptcy in paradox by appropriate amendment or expansion of its premises. This stratagem, of course, engenders new paradox. But now logic adopts the pragmatic attitude; it looks away from first things and forward to last things, results and consequences. Bridgman's dictum is strikingly confirmed: We cannot find the necessary pattern of the external world inside our heads. Science appears the more plainly a "continuation of common sense."

One might now be moved to ask: How does science get ahead of common sense? Quine responds:

The answer, in a word, is "system." The scientist introduces system into his quest and scrutiny of evidence. System, moreover, dictates the scientist's hypotheses themselves: Those are most welcome which are seen to conduce most simplicity in the overall theory. Predictions, once they have been deduced from hypotheses, are subject to the discipline of evidence in turn; but the hypotheses have, at the time of hypothesis, only the considerations of systematic simplicity to recommend them. Insofar, simplicity itself— in some sense of this difficult term—counts as a kind of evidence; and scientists have indeed long tended to look upon the simpler of two hypotheses as not merely more likable, but more likely. Let it not be supposed, however, that we have found at last a type of evidence that is acceptable to science and foreign to common sense. On the contrary, the favoring of the seemingly simpler hypothesis is a lay habit carried over into science. The quest of systematic simplicity seems peculiarly scientific in spirit only because science is what it issues in.

We have all—to the extent that we have ever faced experience objectively and got our thinking rationally in accord with it—been doing science all our lives. Plainly this activity bears upon the ends as well as on the means of our existence. It has won for each of us whatever independence we exercise as citizens. If we can ignore, for a moment, the celebrated technological and economic consequences of the scientific enterprise, we may join Percy Bridgman in wondering what will be the effect of "letting loose in society an increasing quotient of intellectual integrity." At this hour—even as we contemplate the lunatic prospect of our self-destruction by the power science has placed at our disposal—we can envision the possibility that all men may yet hold full membership in the human race, each man to find the opportunity to fulfill the unique potential of his mortal life.

On Recollection

THE doctrine of special creation clings tenaciously to its last niche in the widespread popular anxiety excited by the thought that a computer might think. A gathering of the friends of the Smith College Library in April, 1968, seemed an appropriate occasion to try to put both people and computers in their place.

I F EVER in our species ontogeny recapitulates phylogeny, it is in the college library. That is where the next attempt to realize our humanity begins. Each student finds a unique path through the labyrinth of the stacks to make some portion of human experience his own. Out of such recollection of the past springs the new trajectory into the future.

Should we not, however, be getting on with the retirement of libraries? Gutenberg, we are told, has been reinterred without tears. In place of linear learning from words set end-to-end, we are to have instantaneous information transfer, with audio and, perhaps, tactile and olfactory stimuli to reinforce the visual. The knowledge industry has materialized just in time to absorb the knowledge explosion. Libraries can no longer contain the printed record. The world's greatest university library is about to add a subterranean million cubic feet to its capacity, and there is doubt that this will serve the needs of the next ten or twenty years. With a catalog already as big as a college library, how can the Harvard Library go on storing and retrieving the expanding stock of knowledge? Has the time not come to replace the library with the computer?

On the old-fashioned principle that it is well to begin discussion with a definition, let us be clear about what a computer is. The best description I know of was supplied by the physicist Louis Ridenour away back in 1952 when this momentous development in technology was just getting under way.

Ridenour said: "They are called computers, simply because computation is the only significant job that has so far been given to them. The name has somewhat obscured the fact that they are capable of much greater generality. . . . To describe its potentialities, the computer needs a new name. Perhaps as good a name as any is 'information machine.' "

In the few brief years since these words were written, the computer has, in fact, begun to demonstrate its generality. It is already, in certain fields of science, making significant contribution to the task of containing the accelerating expansion of the literature. But information technology is not progressing as fast and as fruitfully on this line as its enthusiasts hoped for and predicted a few years ago. Progress is constrained for the present not by the capabilities of the technology but by limitations on our understanding of the function we want it to serve. The effort to bring the computer into the library has inspired novel inquiry in library science. The necessary understanding continues to elude inquiry, however, because the function of the library in teaching and learning, in scholarship and discovery proves to be closely related to the deepest questions in information technology. Those questions in turn bear upon the nature of the thinking process itself.

What the computer must replace, it turns out, is not the library but the librarian!

As any librarian can tell you, the operations of a library, on a man-hour basis, consist chiefly of the kind of clerical work that employs the largest single group of workers in our industrial economy. Computers are taking over those functions from clerks in every kind of business, and they are doing the same thing in big libraries. If a computer can write purchase orders, it can buy books. If it can bill customers, it can dun forgetful borrowers. If it can keep asset accounts, it can maintain continuous inventory of the library shelves. If it can sort payments from the random order of the mails, it can sort a catalog into alphabetical or other order.

Information technology is coming forward, moreover, with a range of solutions for the special library problem of storage space. Microfilm, microfiche, and videotape, respectively, reduce 45, 125, and 350 unit-documents (8.5×11 inches) to a single cubic inch. Microminiaturization, employing monochromatic laser light and high-polymer rather than metal salt emulsions, can reduce the unit document to one ten-thousandth of its original size. The Library of Congress presently settles for reduction to one four-hundredth in the demonstrated technology of economically viable systems.

Capacity to store must be matched, of course, by ability to retrieve. Aperture card systems can scan 33 documents per second. One videotape system scans 1,050 documents per second. That system has been demonstrating its capacity in the management, inevitably, of the data bank of a major installation of the National Aeronautical and Space Administration.

Yet, for all these developments, libraries are still acquiring books. More than fifty academic libraries in this country contain one million books or more. The burgeoning of new collections that must reach backward into the past as well as keep pace with new publication has evoked a new sub-industry in book publishing: the reprinting of books long out of print.

To one of these publishers I am indebted for my copy of that seminal work *On the Economy of Manufactures and Machinery* by Charles Babbage. In the 1830's, Babbage ruined his already bilious disposition in the attempt to build a working model of his design for an "Analytical Engine." Historians of science now recognize that Babbage's conception comprehended the essential elements of the computer in the full contemporary meaning of that word.

In its present realization the computer is not yet of much help to the librarian and is still a long way from replacing him. The difficulties are presented, as I have said, not by the hardware but by the function the hardware is to serve. That is: to help the inquirer to find the book or the document he needs. The problem to be solved here is a paradox. Like so many paradoxes, it can be stated innocently. A library is collected and organized in anticipation of its users' needs. As Ben Ami Lipetz of the Sterling Library at Yale has put it, a library must be ready to answer questions before they have been asked. The paradox is this: To the degree that the library has been organized to respond to known or anticipatorily defined needs, to that degree it must tend to fail and to frustrate unanticipated needs.

Librarians know this paradox. Good librarians try constantly to make their libraries respond to the changing needs of their constituencies. They know they are failing when, in the words of Lipetz, "unusual demands on their systems diminish and users take their problems elsewhere." Thus, at Radcliffe, we know there will always be some girls who will exchange the convenience and comfort of the beautiful new Hilles Library for the secure sense of access to all knowledge a half-mile away at Widener. It is one thing to index a NASA data bank and quite another to collect and catalog a college library against the undergraduate's propensity to ask new questions.

It is apparent that hardware cannot resolve the paradox of the library. Without deeper understanding of the ways in which people form associations of ideas, it is dangerous to contemplate the commitment of the library to the power of information technology. The worst thing a computer can do to you is to do what it is designed to do; that is: to do precisely what you tell it to do.

The frontier questions in computer design, therefore, are questions about the uniquely human faculty of thought. To ask whether a computer "really thinks" can perhaps clarify one's thinking about thinking. A. M. Turing—the brilliant young British mathematician who staked out the formal theory of the computer as a universal machine before World War II—proposed a conclusive experiment. Suppose, he said, that the experimenter sits in one room and poses his questions, and suppose the answers come from either a man or a computer in another room. Then, suppose the experimenter cannot tell, from the nature of the answer, whether it comes from a man or a computer. Must we say the computer is computing and not thinking?

Quite rudimentary computers can be programmed to play tic-tac-toe to win or draw. Bigger computers play championship checkers and still more capacious ones play formidable but stereotyped chess. For the future, the computer scientist Marvin Minsky observes: "It is unreasonable to think machines might become nearly as intelligent as we are and then stop. . . . Whether [in that case] we could retain some control of the machines, assuming we would want to, the nature of our aspirations would be changed utterly by the presence on earth of intellectually superior beings."

The computer presents an instructive model of the human brain that made it. A computer is not to be confused with a calculating machine. The difference between these two very different kinds of machine is germane to this consideration of the task of recollection. The computer has a memory. Its memory, in fact, is most of what a computer is. The memory in consideration here is not that contained in the external tape drives, drums, and disks with which the computer conducts discourse through its input and output channels. It is the computer's so-called internal memory that makes up most of its cubic bulk; in general, the bigger the computer, the more of it is memory. In a big machine, the logic operations engage perhaps 100,000 circuits; while the internal memory occupies 5 million to 10 million circuits. This memory is directly involved in the operation of the machine; it is, in a sense, the site of that operation.

The active internal memory holds not only the data summoned into operation from the external memory, but also the program that tells the computer what to do. One of Babbage's special inspirations is routinely demonstrated in the modern computer: It can call in a new program from its external memory when the resident program encounters an obstacle. This key principle in the structure of the computer gives it the power to satisfy Turing's definition of the universal machine: a machine that can do any operation for which a program can be written. Thus, these machines serve variously the functions of master tool- and die-maker, open-hearth furnace operator, space-vehicle navigator, and insistent and responsive mathematics teacher; or as a supermicroscope for inspecting the structure of molecules and as a working model of the economy capable of simulating the consequences of an outbreak of world peace. Programs can be and are written to call in other programs and to modify their own routines in accord with experience and so to approach generality in practice as well as in theory.

The analogy of computer to brain extends to anatomy and physiology. The computer memory may be said to correspond to the great hemispheres of the brain, which are themselves the locus of memory. But the hemispheres contain the "programs" as well. The old hind brain serves as a subsidiary switchboard, mediating input and output traffic with the rest of the organism and with the environment.

This anatomical and physiological comparison gains force when one considers how rapidly over the past twenty-five years the computer has been approaching the brain in size. The computer is constructed of elements that embody knowledge of the physics of the solid state gained since the end of World War II. In theory, the same circuits could have been built around pre-war vacuum tubes. Such a machine, of middling contemporary capacity, would have occupied the cubic footage of the Smith College Library—and it would have required the Connecticut River to cool it! Present-day solid-state technology can pack one million circuits in a cubic foot. According to John von Neumann—who elaborated Turing's insights into a systematic mathematical theory of automata—the logic and memory circuits are packed in the human brain to a density of one million-million circuits per cubic foot. Computer design has thus a million times its present sophistication still ahead of it. But it has gained one thousand times upon the limits of only thirty years ago.

There remains to consider now a still deeper correspondence

between brain and computer. The computer can hold a vast diversity and volume of information in its memory because all information can be coded in that pair of ultimate abstract symbols: 1 and 0. In the computer they are encoded, in turn, in the two states of a switch, either on or off, or in the two states of a magnet, pointing either north–south or south–north. The brain similarly owes its enormous memory to its capacity to encode information in symbols. Words are, of course, the principal symbols. We load words with so much meaning that it takes an effort to recall that these arbitrary symbols are made up of meaningless sounds (phonemes) and acquire meaning only when combined by convention into words (morphemes).

The remarkable invention of language has been described by Roman Jacobson of Harvard University as the "original machine tool." From the first branching-off of our species, language and tool-making played a dynamic role in the natural selection that elaborated the great hemispheres to contain the faculty of memory. Only in comparatively recent times did the artifice of the written word make possible the amplification of that faculty in the external memory of the library.

The brain has now begun to make a model of itself in the computer. To ask this machine, at this early stage in its development, to operate a library is to ask it to facilitate the primary task of recollection. In all the other functions it already performs so well—in operating a machine tool or teaching arithmetic—the thinking has already been done or consists in exploring and exploiting what is contained in the premises. In the task of recollection, the thinking has just begun. It is the task of recollection to subserve the creative act that brings new experience into meaningful relationship with the past. The library holds not a mere collection of information, not a gas of facts, but the tissue of understanding that makes experience make sense. Inquiry does more than add to the record; it changes our understanding of it. In library language, the student, the scholar, and the scientist work against the catalog and force its constant revision. The next great task for the computer is to make the library responsive to this innovative enterprise. The external memory of our species must serve as a source and not a sink of recollection.

King Solomon's Ant

FROM the work of T. C. Schneirla, the subject of the essay that follows, I would here draw three explicit lessons. Schneirla and his work give the lie, in the first place, to the misconception of science as a kind of fact-collecting that can be conducted by a man without quality. He worked in a field—comparative psychology—that is rife with controversy, to which he contributed his share, and rich in human interest, which he warmly embraced. His passion was at all times, however, under prior commitment to the Puritan ethic of knowledge. Animal behavior is the subject matter of comparative psychology. It sets traps at every hand for the unwary, the sentimental, the seeker after final causes; it offers wide open spaces to nature-fakers, soothsayers, and charismatics. Schneirla left the field with compelling models of the interplay between hypothesis and the test of evidence that give it an inductive method of its own. He was always a formidable antagonist in controversy because he had put his own and his opponent's work under the same unsparing criticism.

Schneirla made superlatively productive use of his freedom to inquire. No one else would have thought to choose, and there was no inducement but his own curiosity to lead him to choose, the army ants as the centerpiece of his lifework. Other animals were more fashionable and more accessible. He picked these denizens of the tropical rain forests because they presented so many interesting and difficult questions. None of these questions was as significant as those he had to ask as soon as he got to know the animals at closer range. The result of his work is not a new insecticide, a new medicine, or a new antipersonnel weapon but a hinge on which the comparative study of animal behavior will turn for a long time to come.

The third lesson has to do with the institutions through which society fosters the freedom of such men to do their work. During the best years of his life Schneirla was curator of animal behavior at the American Museum of Natural History. This was an ideal post not because it relieved him from teaching obligations; as a teacher, with a concurrent appointment at New York University, Schneirla created a movement in comparative psychology identified with his tough-minded methods. The museum provided the right setting because natural history is concerned with the whole animal. Departments of psychology in universities and medical schools understandably take the motivation for their research from the professions for which they are preparing their students; they tend to be interested only in aspects of animal behavior that have been shown to be relevant to the professions they serve. Schneirla's work on whole animals and complete life histories shows how such constraints can take good men down false trails; it suggests further how new fields of learning and new professions come to be established. The degree of freedom actually open to creative scholars and scientists will necessarily correspond to the diversity of opportunity and resources that society affords.

This essay was written for a Festshrift put together by students of T. C. Schneirla and published (Development and Evolution of Behavior, edited by Lester Aronson et al., W. H. Freeman and Co., San Francisco, 1970) in his memory after his untimely death in 1968.

AMONG the original contributions of T. C. Schneirla to comparative psychology must be counted his studies of the behavior of students of animal behavior. He was impelled to this uninviting task by his concerned realization that behavior is everyman's science. From Aesop and Solomon to Pliny and Aquinas and, in recent times, from Fabre, Maeterlinck, and Seton to Konrad Lorenz, Robert Ardrey, and Desmond Morris, authors have edified their fellow-men with animal stories. It is now possible to point to some seven decades "of real progress in [the] materialistic and empirical study of animal capacities." At the same time, this work has not much raised the quality of popular discourse.

Anthropomorphism no longer supplies the ready-made rationale, but the contemporary fallacy of attributing the faculty of the brute to man is no less pathetic. The mainstream of comparative psychology has supplied cant along with certification to recent popular literature. Terms such as "territoriality," "pecking order," "aggression," "imprinting," and "reinforcement" mark the trail from the newsstand back to the journal of primary publication. The rise in popular interest in comparative psychology reflects the ascendance of the behavioral sciences in scale and scope of activity and in the expectations of society. As a primary discipline of comparative psychology the comparative study of animal behavior has had need of a discerning and responsible critic. Schneirla filled that need.

Two centuries ago, the French polymath Réaumur could declare with satisfaction: "What the erudite of former times seriously proclaimed to other savants would today scarcely be recounted by credulous nurses to their nurselings." Yet, as recently as 1949, Schneirla found it necessary to remark on the propensity of colleagues to "generously endow" lower animals with capacities "not readily acquired by man." The journals of behavior still carried echoes of the quaint conflict between sentimental vitalists and vulgar mechanists that had died out in other branches of biology a half century earlier. On the one hand, Schneirla could cite a fully articled contemporary who held that psychology was a science of "psychic realities," which man first discovers "by introspection" in himself and then studies in lower animals through "sympathetic intuition"; on the other, he had to confront the author of the empty declaration that "the raw basis from which [a man's interests and desires] are developed is found in the phenomena of living matter." On their divergent courses, these two writers converge in the same self-defeating "reductionism." Each reduces the psychological capacities of animals to a single level, and therefore, apart from other flaws in their viewpoints, they miss the true center of interest in comparative psychology, which is, in Schneirla's words, "the nature of the differences among levels of behavioral capacity."

One may ask why issues so sterile and works of such little merit should claim the attention of a competent scientist with his own work to do. Schneirla made the same use of this material that medicine makes of gross pathology: exposure and examination of overt symptoms sharpen the diagnosis of the covert symptoms of a disease. Animal psychology has not outgrown naïve anthropomorphism when serious investigators can blur the distinction between adaptive and purposive behavior. Schneirla's principal predecessor in field studies of the army ant had, for example, explained the migratory habits of these animals as a response to the depletion of food in the nesting area. "But there is absolutely no factual justification," Schneirla took great pains to show, "for the conclusion that the adaptiveness of such group behavior accounts in any way for the events in individual behavior through which it appears." Evidence of more subtle contamination of the thinking process along the same lines caused Schneirla to urge students of invertebrate learning to recognize that their publications showed "a strong tendency to impose the mammalian pattern deductively upon the performances of lower organisms and to interpret them as simpler editions of that pattern. . . . Apparently, goal-striving is seen as dominating the special process of learning in maze-adjustment on all levels."

The "spirit of the hive," evoked by Maeterlinck to explain the behavior in the social bee, finds reincarnation in the concept of the "superorganism," a notion that has bemused a number of the ablest field-working psychologists of this century. The protagonists of the superorganism concept invariably draw it forth from their astonished cataloging of parallels they have perceived between the anatomy and physiology of multicelled organisms and the organization and functioning of societies, both insect and human. Here Alfred E. Emerson, the most distinguished recent advocate of the superorganism, finds convergent evolution at work: "The common characteristics of organisms and population systems are the result of natural selection of each unit through its internal and external relations." Here, he finds deep principles to be discovered: "Significant parallels between insect and human societies have a common causation." Here, important moral lessons: "Just as the internal controlled environment of the organism is exploited by parasites, social parasites have risen to exploit the social environment."

In pinning down this specimen in his collection, Schneirla observed, as to the integrity of the concept, ". . . individual [organism] and social group are entities in very different senses of the word, differently subject to selection in evolution." As to the method, he said: "While . . . analogy has an important place in scientific theory, its usefulness must be considered introductory to a comparative study in which differences may well be discovered which require a reinterpretation of the similarities first noted."

Schneirla could be moved to use a sharper pen. On the employment of "analogy at the expense of analysis," he wrote: "The term 'mind' suggests close qualitative similarity wherever it is used, evading explanation but at the same time implying it with such propagandistic success that the issue seems fairly settled." On the superorganismic notion of "crowd mind," he wrote: "Such concepts implying a social psychic agency either are used frankly as literary conceits, or as fraudulent devices substituted for any real effort to study the factors underlying a given level of group performance." On the associated practice of anecdotalism: "One who sets out to demonstrate that proto-zoan organisms or any others have the mental characteristics of man may convince himself at least, provided he singles out opportunely the brief episodes which seem describable as instances of perception of danger, of reasoning, or what not. By the same method, the absence of reasoning in man can be proved with ease."

In what other field of science can one today find serious workers arguing from analogy, relying upon anecdote, or pursuing the circular error of nominalism? Such weakness of intellection underlies the recurring "heredity-environment" ("nature-nurture," "instinct-learning," "innate-acquired," "preformistic-epigenetic") disputation that fills the literature. Nothing more insightful or penetrating distinguishes the approaches of the two schools that held the center of the stage in animal psychology during Schneirla's lifetime: the operant-behaviorists identified with B. F. Skinner and the ethologists identified with Konrad Lorenz.

The central concept of ethology is, of course, the "ethogram" or innate behavioral trait, which is programmed in the genes, faithfully unfolded by growth, and triggered into function upon presentation of its appropriate "releaser." Schneirla argued against this notion, using evidence from physiological genetics: "Genes are complex biochemical systems, integrated from the beginning of ontogeny into processes of increasing complexity and scope, the ensuing progressive processes always intimately influenced by forces acting from the developmental context. . . . It is undesirable to confuse this natural state of affairs with the impression of heredity directly determining development." Further, he argued, "There seem to be no hard and fast rules for distinguishing hypothetically innate behavior from other kinds." Pointing out that the ethologists' preoccupation with the innate-acquired dichotomy caused them to focus on part-processes of behavior in the adult animal, Schneirla asked, "In what particular stage of development is 'original nature' to be identified?"

Schneirla commented on the "releaser," citing a well-known experiment of N. Tinbergen. An airplane-shaped model of a "bird of prey" was said to release the appropriate alarm and escape behavior from young nestling birds when it was moved over them with the "shortneck" end first but not when moved "tail first." Schneirla proposed a simpler explanation for this result—the initial effect on the birds might be "perceptually non-qualitative, perhaps a shock reaction produced by a sufficiently abrupt stimulus." He was able to cite an experiment by another worker that showed the speed of the moving effigy to be the critical factor in eliciting the response of the nestling. "The artifact method," he concluded, "requires careful control against subjective impressions. . . . 'Shortneckedness' in the above situation is a cue to the human observer . . . but not necessarily to the bird."

Schneirla found the conceptual poverty of the notion of the innate behavioral trait plainly exhibited in the kind of experimentation it inspires. By way of extreme example, he pointed to efforts by various investigators to demonstrate "information transfer" by the injection or feeding of extracts of genetic material, which the public has come to know by the initials DNA and RNA, from animals trained to perform specific tasks into the eggs, embryos, and mature specimens of untrained individuals. The ultimate enterprise in this direction is represented by the planting of electrodes in the presumed "pleasure center" of the brains of catatonic patients in a Louisiana state mental hospital.

The operant-behaviorists approach their subjects from a quite different quarter. Skinner declared that he and his colleagues had been able to demonstrate "surprisingly similar performances, particularly under complex schedules in organisms as diverse as the pigeon, mouse, rat, dog, cat and monkey." This work, said Schneirla, "is directed at just one set of abilities in the area of learning: those having to do with the establishment, weakening by extinction, re-establishment, and so on, of continuity-learning (i.e.: Pavlovian) patterns." It is so contrived "as to largely exclude or obscure the animal's potentialities for behavioral organization." Schneirla noted that these behaviorists "have not claimed to offer a comprehensive psychological theory," and he pointed to the hazards "in the uncritical extrapolation of [their] results to the broader problems of comparative behavior." In another connection Schneirla identified one such hazard: "When an educational program is unscientific in that it emphasizes a mass acquisition of predigested categorical results and discourages questioning and search for better answers to old questions, the conditioned-response pattern of learning obviously is in the ascendance and the intelligence potential of the given society is broadly wasted."

Schneirla wryly took note of a conference of ethologists and operant-behaviorists called to explore the possibilities of collaboration from their respective sides of the heredity-environment dichotomy. They found some grounds of common interest but believed the possibilities of collaboration to be very limited. Schneirla thought this conclusion was unduly negative. The two parties "may be more closely related in their basic rationales than this conclusion might suggest," he observed; both groups rely on "implicit or outright assumptions of innate behavior as distinct from learned behavior." Schneirla then added: "Behavior develops and changes—on higher integrative levels it is not often rigid except when pathological. Yet ideologies directed at interpreting it are too often cast in rigid forms."

It is a fact deserving a moment of reflection that the instincts and releasers, the stimuli, reinforcements, and responses, the superorganisms and the very animals themselves featured in the contemporary literature of comparative psychology hold less inherent interest than, say, the time, space, particles, waves, forces, and quanta of modern physics. Yet physics stands at the "disorganized simplicity" bottom of the hierarchy of the sciences postulated by Warren Weaver, and behavior surely belongs to the "organized complexity" category at the top. In responding to the question why behavioral science should be so much less prepossessing than its subject matter, a behavioral scientist is obliged to give some credence to surveys that show physics has been attracting more able intellects, as identified by various tests. There is another answer, perhaps kinder to the practitioners. Most of them have other primary interests. They are taxonomists. They are concerned with range and wildlife management. Or they select animals as models for study, in other contexts, of perceptual and learning capacity or of straightforward neural physiology. Not many have been concerned with the work of comparative psychology as such. Not one in recent years has done more to define and to keep forward the significant questions in this field than Schneirla.

"Theories concerning how behavior and behavior patterns arise," Schneirla wrote, ". . . must do more than bridge initial and terminal stages with hypothetical shielded intra-organismic determiners." The gap "between genes and behavior" is not to be bridged; it is the territory to be explored. Schneirla remembered always that life is history and that ontogeny is therefore the proper study of comparative psychology. "Processes of behavioral organization and motivation cannot be dated from any one stage, including birth, as each stage of ontogeny constitutes the animal's 'nature' at that juncture and is essential for the changing and expanding accomplishments of succeeding phases."

The animal of Schneirla's studies is thus a productively different kind of animal from the ethologist's wind-up toy, awaiting its releaser, or from the operant-behaviorist's feathered and furry computers, so susceptible to programming and readout.

Behavior, or for that matter, development, can be seen at any given stage to reflect factors that are internal and external to the animal. This useful preliminary analysis invariably hardens into the familiar dichotomy of heredity and environment. On more than one occasion Schneirla urged his colleagues to extricate themselves from the "heredity-environment" trap and tried to show them how. Starting with the semantically most neutral pair of terms corresponding to the preliminary analysis, Schneirla suggested that " 'maturation' be redefined to refer to the contributions to development from growth and tissue differentiation, together with their organic and functional trace effects surviving from earlier development" and that " 'experience' be defined as the contributions to development of the effects of stimulation from all available sources (external and internal) including their functional trace effects surviving from earlier development." These definitions, it should be observed, make a necessary correction of the simplistic distinction drawn between the external and the internal. Influences, both external and internal, become incorporated in the trace effects of progressive development; through feedback linkage, the output of earlier stages of development becomes input to the later stages. In consequence, Schneirla concludes: "The developmental contributions of the two complexes, maturation and experience, must be viewed as fused (i.e.: as inseparably coalesced) at all stages in the ontogenesis of any organism."

From this operation, the heredity-environment dilemma departs without its horns. Consider, for example, the familiar "isolation" experiment designed to establish the "innateness" of some behavioral trait; "proof" is furnished if the subject exhibits the trait upon release from isolation. In Schneirla's words:

Isolation experiments do not tell us what is native in the normal patterns; for, if the animal survives, the atypical situation also must have contributed to the development of some adaptive pattern. Techniques of the "isolation" type thus concern relative-abnormality-of-setting rather than isolation in the full sense, and help tell us how far extrinsic conditions may be changed at a particular developmental stage without preventing or altering further development based on the gain from preceding stages.

Study of the ontogeny of behavior sets the only secure foundation for comparative psychology—for "comparing the respectively different adaptive patterns attained by animals and for better understanding of the animal series inclusively." Quoting Z.-Y. Kuo, Schneirla pointed out that such study answers not only the question "What happens?" but "How?" Without an adequate longitudinal perspective on the developmental process, the investigator is prone to error when he compares analogous or homologous patterns of behavior observed in unrelated species. The literature of animal behavior is rich in misconception, as Schneirla's periodical surveys and critiques have shown. Misconception is reinforced by cross-sectional or time-limited studies that often "turn out to be demonstrations of the investigator's initial assumptions rather than critical tests of the theory that inspired the work."

Investigation of traits, habits, or modes of behavior isolated from the context of the subject's development is inappropriate for comparative studies because behavioral evolution cannot be viewed "as a process of accretion" with new mechanisms added to old. "Homologous mechanisms are transformed functionally . . . as through ontogeny a characteristically total behavior pattern arises in a different total context."

This observation has particular relevance at the present time. The public is being told, through popularizations of the scientific literature, that the stress of history is uncovering man's dark genesis and stripping the veneer of culture from the naked brute. It is not the brute, Schneirla insisted, but a pathological human personality that is beheld on such occasions: "The view that man's 'higher psychological processes' constitute a single agency or unity which is capable of being sloughed off under hypothetically extreme provocation is a naïve outcome of the mind-body conception of man's nature."

As a common denominator for comparing behavioral capacities attained at successive levels of evolutionary development, Schneirla proposed a general theory of "biphasic processes underlying approach and withdrawal." He submitted, with italics, the statement that: "*Approach* and *withdrawal* are the *only* empirical, objective terms applicable to *all* motivated behavior in all animals." He was careful to specify that "an animal may be said to *approach* a stimulus source when it responds by coming nearer to that source, to *withdraw* when it increases its distance from the source."

Yet he had to admit: "This point is not sufficiently elementary to escape confusion." Some people find it easier to oppose "approach" with "avoid," the source of their confusion being, of course, their inexplicit identification of adaptation with purpose. With teleology set aside, however, it is plain to see that approach and withdrawal behavior must be highly adaptive in all surviving animals, for "beginning in the primitive scintilla many millions of years ago . . . the haunts and typical niche of any organism must depend on what conditions it approaches and what it moves away from." Of the biphasic response it may be said, in general, that "low intensities of stimulation tend to evoke approach reactions; high intensities, withdrawal reactions."

Compared to the imaginative hypotheses that inspire so much current investigation in animal behavior, the "approach-withdrawal response" appears as a pallid truism. This quality nicely served the purpose of the Schneirla program: to develop in an unambiguous way the differences in the ontogeny of a simple, universal response, from species to species.

In the amoeba, approach or withdrawal is energized by "protoplasmic processes set off directly by the stimulus"—the sol reaction in response to a weak source of light, with streaming in the direction of the source, and the gel reaction in response to a strong source, with streaming in the opposite direction. In the rat, the response is mediated by "specialized higher-level processes not indicated in the protozoan"—including extensor muscle dominance with weak stimulation and flexor dominance with strong (as had been demonstrated by the great British physiologist Charles Sherrington)—and is also freed by ontogeny from the sway of simple stimulative intensity. The human infant, Schneirla observed, "specializes perceptually in reaching and smiling before he avoids and sulks discriminately."

The Schneirla program for comparative study of the approach-withdrawal response poses questions in a fruitful way for investigations yet to be mounted. It also raises serious questions about the kinds of answers that have come from past studies— answers that have given the sanction of science to the popular hankering for final causes. One set of finalisms is represented by the various constellations of emotions and drives with which human beings are supposed to be endowed at birth. J. B. Watson, the founder of the American school of behaviorism, postulated a trinity of innate emotions—"love," "fear," and "rage"—from his observations of the newborn human infant. As Schneirla suggested, further study will doubtless show that, in the psychosocially barren dawn of human existence, these come down to a simple approach response for the first and withdrawal for the second two. Watson himself regretted that he had attached emotion-laden words to the responses he observed and expressed the wish that he had designated them "x," "y," and "z" respectively. In that case, Schneirla said, "x" would reflect extensor dominance and "y-z" tensor dominance. As for the pleasure-pain dichotomy, now ratified by the identification of the "pleasure center" in the brain, it may yet be shown that the implanted electrodes "cut-in on A- or W-type patterns, tapping critical way-stations in circuits of distinctly different arousal thresholds."

If it can be said that the science of behavior has approached complete elucidation of the behavioral repertory of any animal, this surely must be said of Schneirla's work on the army ant. The questions that remain after Schneirla, that indeed were posed by Schneirla, must be answered through the methods of other disciplines: neurophysiology and hormonal physiology, to begin with. In this work, he not only set a model for his science —of a system of behavior mapped and structured in minute detail—but he also showed that behavioral science can develop methods and concepts worthy of its subject matter.

Schneirla chose as the centerpiece of his lifework "the most complex instance of organized mass behavior occurring regularly outside the homesite in any insect or, for that matter, any infra-human animal." This is a sober understatement of a phenomenon that had caused earlier students to abandon discretion. Describing an ant army on the march, Paul Griswold Howes in his treatise *Insect Behavior*, published in 1919, has the queen "hidden from the common horde, attended by her special ladies-in-waiting"; eggs, larvae, and pupae "guarded and kept warm, lest injury result and the future of the tribe be endangered"; "lieutenants keeping order or searching for members out of step that might hinder the march"; "scouts searching out the ground to be hunted or travelled next"; and the run-of-the-mill ants "obeying commands" and evincing a "wonderful sense of duty." T. Belt, a nineteenth-century naturalist who came from Britain to observe the army ants of Central America, noted that it is the "light-colored officers" that keep the "common dark-colored workers" in line.

Schneirla's observations and his experiments with individual colony members showed that the army ant is a highly limited animal, with less capacity for maze adjustment than the common trail-running *Formica*. Plainly, the organization of colony behavior "does not pre-exist in any one type of individual workers, brood or queen—nor is it additive from these alone."

As Schneirla showed, the system arises from the intricate, inter-dependent interaction of a large number of diverse factors. They may be conveniently classified under the headings of individual colony-member capacities, group functions, repro-duction, the development of the brood (eggs, larvae, pupae, and callows), the natural environment, and the system of organized behavior itself. There are many cross-links between behavioral subroutines; a hierarchy of interlocked feedback loops integrates the total performance. Thus, for example, the mass of living workers in the colony bivouac ensures that "a stable microclimate is so regularly established for the successive broods that the periodicity of the developmental stages is highly predictable in brood after brood." But the developmental cycle of the brood plays its part in turn: "Reciprocally, through indi-rectly contributing to the existence of rhythmic colony behav-ior, a normally developing brood makes the bivouacs possible."

The biphasic approach-withdrawal response is to be seen in action at the head of a raiding swarm or column, with indi-viduals venturing forward, beyond the "trophallactic" thrall of the characteristic colony odor, and rebounding back into the mass when that stimulus has been weakened by distance. In the laboratory and in the field, Schneirla recorded episodes showing that the stereotyped army ant workers, when they are placed in neutral surroundings, such as a tabletop, a sidewalk, or a flat rock, inexorably form up in a circular mill and march them-selves to death. It follows that "heterogeneity in the operating terrain is indispensable for execution of the typical raiding-emigration sequence."

Considering the entire picture in all its complexity, it is tempting to attribute an endogenous control of the cycle to the queen "who becomes physogastric and delivers a new brood approximately midway in each statory phase." Yet, on closer inspection, Schneirla found that the queen's periodicity is regu-lated in turn by the maturation of the larval brood. As the brood approaches pupation and reduces its feeding demands, the surplus of food and worker activity is diverted to the queen. The colony is meanwhile terminating the nomadic phase of its migratory cycle, and the queen, feeding voraciously, enters her next ovulation cycle. "The cyclic pattern thus is self re-aroused in a feedback fashion, the product of a reciprocal relationship between queen and colony functions, not of a timing mechanism endogenous to the queen."

This cyclic drama, which has been reenacted in the tropical rain forests of the middle latitudes of the planet week in and week out for tens of millions of years, is not to be reduced to any single internal or external cause. Nor is it to be explained by any governing or energizing principle extrinsic to its own dynamic organization. Schneirla's careful analysis and reconstruction show the army ant behavioral system to be *sui generis* and as splendid, in its fashion, as a supernova.

Schneirla was intensely aware of the liabilities that an investigator takes with him into the field or laboratory. "Man goes to nature," he said, "to learn what nature is but, in so doing, he introduces possibilities of distortion through his own presence." He warned of a second hazard: "When initial concepts are strong and vivid and metaphorically very appealing, they may be carried through an entire study without ever being examined effectively." With reference to the special hazards attending the study of behavior, he declared, "Our attitudes toward the nature, origin and relationship of competition, cooperation and natural selection processes exert subtle influences not readily controlled in the planning and prosecution of investigations." Whether in field or laboratory, "the experimenter's responsibility for control involves the regulation of such matters, and not only the manipulation of objective factors in the phenomenon under study."

Befitting the role of the behavioral scientist, Schneirla set out his position not only in precept but also by example: "It is the methods of objective science that must be directed at the question," he said, "and not the forensic arts." He schooled himself to keep detailed, written records of what he observed while he was observing it. With a command of shorthand and a practiced ability to write in the dark, he was able to take notes "by tactual control" while maintaining "visual touch" with events. He also practiced estimating distances, counting animals, and judging their spatial arrangement; he worked out codes and symbols to record the interaction of his subjects. Frequently his notes included on-the-spot sketches, diagrams, graphs, and photographs. Schneirla was careful to record the usual as well as the unusual. At the end of each day's work it was his habit to review and summarize the day's record. At intervals of a few days he would review the accumulated record "for discrepancies, omissions and latent meanings." With conclusions and hypotheses spelled out from earlier observations, he would repeat his observations systematically on the next round of his subject's behavioral cycle.

Many of Schneirla's papers report experiments devised in the field. Thus, by removing the larval brood from a colony, he tested his hypothesis that the squirming larvae stimulate the more energetic raiding of the nomadic phase in the cycle; raiding promptly declined and nomadism soon ceased. He later supplemented this field experiment with laboratory tests of the relative responsiveness of individual workers to active larvae and quiescent pupae.

By such stratagems Schneirla sought to bring to his field work the rigor that is associated with the laboratory. In his view, there is no inherent virtue that makes experiment more reliable than observation. The study of animal behavior is necessarily an observational science. Knowledge from the field is essential, at the very least, to correct for "the influence of the laboratory environment, the cage." The ideal program for the future study of animal behavior, he concluded, "would involve a coordination of field and laboratory investigation. . . . Field investigation offers an opportunity to work with the animal's full pattern of activities. . . . In the laboratory one may focus on specialized problems such as sensory discrimination, motivation, learning and higher processes, pursuing them in detail and under conditions involving refined controls."

Schneirla was a scientist to whom a wise student would report as an apprentice. He was a philosopher of science who practiced what he preached.

III Technology and Democratic Institutions

Why No
Satisfaction?

HINDSIGHT discloses a break in the clouds back in 1963–1964 when I wrote the three essays in this part. Domestic affairs then returned for a while to the center of the country's politics. John F. Kennedy, having campaigned on the nonexistent missile gap, had arrived in office on the wrong foot. By the end of his first year he had pumped the military budget a full 25 percent above the level at which the Eisenhower administration had screwed down the lid eight years before. By January, 1963, however, Kennedy's nervy handling of the Cuban missile crisis had recouped the public and the self-confidence he had lost at the Bay of Pigs and in his panic following his confrontation with Nikita Khrushchev in Vienna later that first year. With this catharsis of his ambitions in foreign affairs he was now budgeting his major energies for attention to the country's household problems. That attention was needed because the business cycle was in a downswing that did not augur well for the campaign of 1964.

Lyndon B. Johnson carried Kennedy's intentions into full commitment promptly upon his accession to office. His proclamation of the Great Society and his mastery of the Capitol cloakrooms promised significant departures in the use of federal power and public funds to brake the turning of the vicious circles in the private sector of the economy. Along with many other citizens I was suppressing recognition of the significance of Tonkin Gulf when I wrote the third essay in this section. In the end, it was not the Great Society but the Indochina war that gave the nation another few years of spurious economic boom and brought us into the 1970's with social inequities as cruel as before and with our domestic public services in still greater disarray.

Now that the public and the not-for-profit private sectors have come to play such a commanding role in our economic life, the relation between technology and democratic institutions deserves more formal study than I have given it here. The relation is reciprocal, and that requires a better understanding of technology among political scientists. Out of the ignorance of technology prevailing among the fellows of the Center for the Study of Democratic Institutions, for whose convocation in January, 1963, in New York City I prepared the first essay in this section, has come a draft for a new U.S. Constitution that has no Bill of Rights—this at a time when technology is enlarging the rights to which every citizen should be entitled at birth.

J UST a generation ago the title of this symposium—"Technology and Democratic Institutions"—would have stood for a roundabout way of saying Progress. In the present American climate it evokes the problems that are gathered under the dispirited heading of Change. The word Change is a gingerly locution for Revolution—not Revolution in the grand old sense of American Revolution, but Revolution as it is beheld from the uncomfortable vantage of a seat in the tumbril.

The alienation of the worker diagnosed by Karl Marx a century ago has given way to the middle-class malaise of anomie, alienation not only from society but from self. Affluence is the terminus of the American Dream, affluence from which, in Edward Albee's transcription of the voice of the people, "You get no satisfaction!" The sky-scraping cities of the land have become slums—as Morton Grodzins and his colleagues at the University of Chicago have shown—transformed by the menacing new pattern of metropolitan segregation. The verdure of the Garden State, says Mason W. Gross, president of Rutgers, the state university, is vanishing under "the creeping, crawling hideousness" of the rubber-tired culture of the suburb. With "an educational system that does not educate and a system of mass communications that does not communicate," in the words of Robert Hutchins, we have become incapable of "the discussion by which political issues are determined." The margin by which we elect a national administration precisely measures the breadth of political discourse. What could come of such an election in these times but an administration impotent in policy and glad to commit its power of decision to a computer?

Americans find themselves looking abroad with unaccustomed envy to the future-dwelling peoples of the newly buoyant economies of Western Europe, with anxiety to the harshly forward-driving socialist systems, with dismay to the new nations of the world's poor, so eager to embrace any system that will bring them to our dubious estate. At home, Americans are looking backward. Some to the thirties, which is what they mean by "Let's get America going again!" Some to the bustling laissez-faire normalcy of the twenties. Many more to the 1860's, in the chauvinistic celebration of the War Between the States that, on January 1 of this year, overlooked the centenary of the Emancipation Proclamation.

The Center for the Study of Democratic Institutions asks us to look back to 1776. We are to answer the question: Can democratic institutions framed in the landscape of a rural republic secure liberty, equality, and the pursuit of happiness in a modern industrial society? That is: whether the individual, whose perfection was to be the goal of our society, is fated now to be the well-fed creature of a technological order that has endowed him with power to no purpose, with techniques for programming and no capacity to plan, with means that compromise all ends.

On these questions the fellows of the Center have been consulting, of course, Plato. The *Republic* has much counsel to offer on the design and management of democratic institutions. It has little to say, however, about economics. An attentive reader will note a passing reference to *andropoda*, the "human-footed" domestic animals. These were not counted in the citizenry, so their presence did not subtract from the prevailing quotient of democracy. Most ancient texts have nothing to say about slavery, not out of embarrassment but because so natural an institution attracted no attention. Jewish codes did call for a minimum humanity in the treatment of slaves but raised no question about the legitimacy of their condition. Later moral philosophers could show that the rights and duties of fief and vassal were founded on natural law. For the most part, authority asked only for the legitimacy conferred by the passage of time.

Economic questions are not so easily avoided today. Scarcity has given way to abundance. The compulsions prescribed by natural law are weakening. Our lives are cast, however, in the uncomfortable transition period. The advent of abundance is not yet comprehended in the theory and practice of our economy. On the contrary, our abundance is dodged, minimized, and concealed as well as squandered and burned. To the degree that we have failed to come to terms with our historic attainment of abundance, we have botched and corrupted its distribution and realization.

The failure to comprehend this failure has sadly misled the students of our plight. They blame the machine for our willful mismanagement of it. This compound failure diverts public discussion from what ought to be the issues before the electorate and reduces our politics to trivia. The stress on our economic arrangements has been transferred to our democratic institutions.

Consider the affluence that yields no satisfaction, the convergence on mediocrity and indistinguishability that characterizes so many of the common articles of commerce—white bread, light beer, and colorless vodka, to mention a few correspondingly tasteless products. This looks so much like the universal process of entropy that it is taken without question to be a bane laid on by technology. The real truth is that technology frees the task of design and production from technical limitations. By doing so, it permits other considerations to hold sway. It is these other considerations, I submit, that assert the common denominator of mediocrity.

Consider the American cigarette. This is the product not of technology but of the new strategy of imperfect competition, or oligopoly. In this kind of competition the two or three competitors all get the same answers from their market surveys and public opinion polls; they look over one another's shoulders in computing the maximization of their results, and so they make their products more and more alike. Now, it can be argued that they make their products this way to take advantage of the economies of mass production; what they save on production, however, they squander on sales cost. In the language of game theory, the cigarette is the solution to a "zero-sum contest with perfect information"; the logic of the situation leads all the players to adopt the same strategy.

The money and the social cost of this way of managing abundance is plainly demonstrated in the American automobile. Here is, above all, a social not a technological artifact, and one symbolic of the pathology to which I am urging your attention. Technological considerations of function, efficiency, and utilization of resources might dictate one or more designs for the automobile. None of the American automobiles is any one of these. The gifts of abundance are bodied forth in waste: of materials, fuel, and space, to name the principal categories. Here also we see the relentless convergence on the same design, the pursuit of identical oscillations in design change, and the packaging that is supposed to make all the difference in the world. Against the argument that mass production must necessarily make the autos all alike, it can be shown again that the sales function integrated in the virtuosity conferred by technology throws away the economies of mass production on junk. As a supposedly durable good, what is more, the automobile does no credit to American technology. According to the standard practice of our durable goods industries—always with the aim of perpetuating scarcity in the face of abundance—the automobile is designed for 1,000 hours of service, to be traded in at 40,000 miles or less.

Now, it is plain that all of this has to do with the maintenance of prosperity, with the creation of jobs, with the struggle for profit margins and other vital objectives. But the compulsions that dictate these objectives arise from the economics of scarcity. They are not determinants of technology. In the place of abundance, proffered by technology, we get affluence.

The Bell Telephone Company, responding to different economic compulsions, builds to far higher standards of service. The telephone handset, the cheapest thing of its kind in the world, is built for amortization over twenty years. During that time, of course, the telephone company sees to it that you put the instrument to a great many more than 1,000 hours of service.

Turning to somewhat larger issues, let me tell you how the late Erwin Wolfson, the author of the Pan American Building, responded to the question: "Who is responsible for ugliness?" The question was put to Mr. Wolfson and others at a gathering hopefully entitled the First Conference on Aesthetic Responsibility, which was held in this city early last year. In the course of his reply, Mr. Wolfson recalled that ". . . the architects came up with a scheme which developed about 1,500,000 square feet on a plot of 151,000 square feet, with a valuation of $20 million." To this, he added: "It just wouldn't work." There was apparently no doubt about this conclusion; for what the architects had to come up with, finally, was a scheme that developed 2,400,000 square feet and a valuation, I am told, of $75 million on that same 151,000-square-foot plot. Mr. Wolfson was able to show, furthermore, that as the result of the sympathetic collaboration of the builder and the architect, the Pan Am Building is not so ugly as it might have been. As it stands, the building is the resultant of a multivariable equation involving a host of economic compulsions and some aesthetic considerations, to the solution of which technology lent high flexibility and freedom.

The one element lacking any weight in those equations was the public interest. But you can look in vain, both north and south on Park Avenue from the Pan Am Building, for evidence that the public interest is represented in the interplay of countervailing forces on which our economists have urged us to rely. The abundance of choices placed at our disposal by technology are harshly narrowed to the solutions of scarcity economics. Technologically speaking, we could build our town the way we like it. Economically speaking—that is, from the obsolete premises of scarcity—it gets built the way no one likes it.

So people flee to the suburbs. The escape is made possible by our rubber-tired transportation industry, and thus by the choice-expanding power of technology. But the whole scheme is as unfair as it can be. It works for the benefit of the better-off at the expense of the worse-off. Reaching out in all directions from every central city in our country there is a radial gradient of ascending incomes. Except for Manhattan and San Francisco and a few enclaves like Louisburg Square in Boston and in other cities, the central cities are becoming racially segregated slums. The nearer suburbs are blighted by the triumphs of highway engineering that bring traffic in at speed from the unspoiled, more distant suburbs.

When we look into the economics of suburbanization, we make a surprising discovery. The public interest is represented there with the biggest expenditures in the public sector after armaments and education. This is the outlay for highways, running at the rate of $10 billion a year—not far, in fact, behind education. What is more, one must reckon a large portion of the education outlay in the economics of suburbanization: the current expenditure per child (for teaching and the like) is down from 1939 (in real dollars), but the investment in the building of schools is up. In other words, the American public has been heavily engaged in financing this transformation of its lifeways. It is a transformation which, by any reckoning, tends to cheat the worse-off, and therefore the larger numbers of people, for the benefit of the better-off. In this case substantial public expenditure has generated no countervailing force in favor of what would seem to be the interests of the larger numbers, if not of the public as a whole. On the contrary, these taxpayers would seem to have financed their own defraudment.

One might wonder how it is that the treasuries of federal, state, and local governments disburse funds so generously for highways. It is not enough to observe that rubber-tired transportation, adding up to a total expense of about $80 billion per year (including the $10 billion for highways), has an economic interest in promoting this capital subsidy and so maintains well-financed lobbies at each node in the governmental network. For the country also has a substantial rail-transportation industry; it once played an appropriately commanding role in our local politics. What is more, it is apparent that electric-powered commuter services could make the shuttle from city to suburb a much less exacting human task, would reduce air pollution and wonderfully relieve congestion in town. The railroad industry has even begun to make half-hearted overtures to the taxpayer. But one gathers, from the Pan Am Building and the impending demolition of Penn Station, that the railroads are less interested in their public franchises than in the real estate they acquired therewith. The airlines, constituting a much smaller element in the transportation sector, already draw much bigger subsidies, for operations as well as for such capital inputs as the design of their aircraft and the building of their airports. If there is anything to the interplay of countervailing forces as a scheme for rational management of society, it is high time our citizenry generated more of them.

The need for ventilation of our political life is becoming increasingly urgent for another reason not often mentioned in public because it leads discussion into such dangerous and trackless territory. Let me take you a little way into the thickets. The dark truth is that our system is ailing. There is even some unanimity about the diagnosis; economists agree that the economy is suffering from a widening gap between its capacity to produce abundance and its ability to generate effective demand in the marketplace. In consequence, for a while at least, the American democracy is urged to bear with an "acceptable" minimum of unemployment, now set at 4 percent.

No one should be surprised by this state of affairs. A technological order that employs fewer and fewer productive workers is obviously qualifying fewer and fewer consumers with the purchasing power to buy the goods it makes. Nor could we want it otherwise; it is, after all, the function of technology to ease man's labors and to multiply his product. It is up to society to make the new arrangements necessary to realize the bounty of technology. In fact, our economic system has managed to do so with more or less success from year to year. Its success is the more remarkable in view of the fact that the latest revolution in technology has been subverting the underlying, inexplicit premises of scarcity. .

Since 1900 the portion of the labor force engaged in productive functions—and these include transportation and construction as well as manufacturing, mining, and farming—has declined from 75 to 45 percent. The decline has come furthest and fastest in farming, of course, and in unskilled labor. At the outset, these declines were offset principally in other lines of production—in the expanding manufacturing sector and its need for skilled labor. Now these demands have leveled and have begun to shrink. Meanwhile our economy has created whole new categories of employment: white-collar jobs in trade and distribution and in the services. I have already remarked on the waste involved in these activities; I must here remark that they serve the vital economic function of qualifying consumers with paychecks and so supplying effective demand for the goods produced by the declining number of productive workers.

The record shows, however, that the system could not absorb the impact of technology without external assistance. Between 1900 and 1929, the percentage of the gross national product that cycled through the public sector, that is through the payrolls of federal, state, and local governments, increased from 3 to 10 percent. Again, whatever other functions these jobs filled, they helped to supply effective demand. The massive income transfer from higher-income taxpayers to lower-income public jobholders greatly expanded the number of consumers qualified with purchasing power. In those days, 60 percent of the public jobs were in state and local governments and were occasioned principally by the rising demand for the welfare and resource-conservation services of government. Lincoln Steffens, you will recall, nicely explained the function of municipal corruption in securing the redistribution of income outside the public payroll proper.

In the 1930's, when the mounting abundance of technology first confronted us with the paradox of poverty in the midst of plenty, national policy took explicit notice of the effective demand that can be generated by public expenditures. Overnight, with popular approval, the public sector doubled its claim on the gross national product, rising from 10 to 20 percent. The ratio of state and local to federal budgets was at that time reversed, and the federal government began disbursing 60 percent of the public funds.

Today more than 25 percent of the gross national product turns over in the public sector. But now there is a difference. Fully 10 of the 25 percent goes to the maintenance and supply of the military establishment. Again, one must observe that whatever other function these expenditures serve—whatever their military justification—they help to maintain effective demand. They also help to maintain the system of private enterprise by providing jobs for corporations as well as people, many of these jobs being functions that used to be conducted by the government, its armed services and arsenals. Above and beyond the call of duty, the war economy has given us affluence.

In helping to solve our fiscal problems, however, the war economy has exacted a measurable social cost. Half of the 10 percent that it turns over has come from increase in the size of the public sector; the other half has come from curtailment of the welfare and resource services of the government. Those services, as a percentage of the gross national product, have been cut by one-third.

The reasons why the military is accorded free access to the public treasury while the claims of our human and natural resources evoke mighty opposition ought to be more closely explored by the students of scarcity economics. Such study is needed because it is apparent that military expenditures must henceforth play a declining role in fiscal policy. Progress in military technology in recent years has drastically reduced the cost of devastation per square mile. The solid-fuel Minuteman and Polaris missiles have made the more labor-intensive B-70 and Skybolt obsolete. Of the ultimate, push-button weapons, as President Kennedy has observed, we can equip ourselves with no more than a finite number. This is a serious situation because the war economy qualifies some 7 million consumers with effective demand. These 7 million plus the 4 or 5 million unemployed—some 20 percent of our labor force—represent the gap between effective demand and our capacity to produce abundance.

The administration now proposes to fill the gap with a tax cut. If this measure helps to restrain demand for further extravagant increase in the military budget, it may serve a purpose. We are assured in advance, of course, that the administration will restrain or curtail other public expenditures. But the question at issue is whether a deficit created by a tax windfall to higher-income groups and business enterprises will generate more jobs. A public-sector deficit averaging $8 billion per year over the last decade has not kept the economy from stagnating. In the present climate there is serious danger that the tax windfall will go into personal savings and corporate surplus accounts and not into consumption and investment. The administration's economists have observed, in another context, that this country has "a backlog of demand for public services comparable in many ways to the backlog of demand for consumer durable goods and housing and producers' plant and equipment at the end of World War II." A deficit created by increased public investment in the human and material resources of the nation would more surely generate employment as well as increase the abundance of our domestic life. Governors and county supervisors would be well advised to seize upon the proposed diminution of federal demand on the taxpayer and turn these funds to better account.

The most urgent need is for increase in the budget for education. Surely an educational system that educates is essential to the operation and perfection of our democratic institutions. With an electorate awakened to the new possibilities—not to mention goals and purposes—that technology has opened up to human life, we might expect to see wiser deployment of our capacity to generate abundance: deployment of that capacity to the creation of more spacious cities, to the cherishing of natural resources, to the attainment of a happier ecological adjustment of Americans to their bountiful environment.

These objectives can no longer be regarded as "residual," to use Charles Frankel's term; they must come to the center of the stage as the primary "institutional" questions of politics and public policy. The public interest cannot, as W. H. Ferry has said, be left to find its way in through the interstices in the web of private interests.

My answer to the question put to us at this assembly is that technology is concerned not alone with the means but also with the ends of life. The toolmaking propensity of man, from his beginnings, has expanded the possibilities of his existence, exciting as well as implementing his aspirations. It is only in the milieu of industrial society that we can speak of equality without the silent discount of eight out of ten of our brothers. It is by toolmaking that men have at least freed themselves from physical bondage to toil and to one another. One of the aspirations that now becomes feasible, therefore, is self-government. With the open acknowledgment that our toolmaking has brought a revolutionary change in man's relationship to nature, we can proceed to the necessary revisions in our relationships with one another.

Science Can Never
Be Retrograde

THE Albert Einstein College of Medicine is one of a dozen "NIH" medical schools set up in business by the huge federal outlays dispensed by the National Institutes of Health from the late 1950's into the 1960's. By advancing the money to support research rather than education, the NIH outflanked the American Medical Association in its opposition to increase in the supply of M.D.'s. "Einstein," established in the academic custody of Yeshiva University in New York City, symbolizes a triumph over opposition of another kind. It assembled a faculty of instant high distinction, made up principally of scientists and physicians who had been denied adequate recognition by the Jewish quota in faculty appointments at class A medical schools or who had made their careers in government health services. By the time the school graduated its first four-year class, for whose commencement this essay was written in May, 1963, it had taken its place among the first five or ten schools in the country.

THE Constitution of the United States nowhere mentions science. The authors of our state papers are never found guilty of such oversight today; they can be relied upon, in fact, to extol not only science but the atom and space as well. We must not, however, be misled by appearances. The truth of the matter is that in eighteenth-century America science went without saying. The authors of the Constitution were natural philosophers; they counted physicians and engineers among their numbers as well as lawyers. Let Thomas Jefferson speak for them.

In 1799, Jefferson wrote a long and discursive letter, in his plain and readable hand, to a young man who had sought the counsel of the founder of the University of Virginia about his proposed course of study there. This letter was first published, after 133 years, in facsimile, in Vol. I, No. 1, of *Scripta Mathematica*—that remarkable journal founded and edited for three decades by the late Jekuthiel Ginsburg, professor of mathematics at your university. After giving the young man some hard-headed advice, Jefferson declared, "It is impossible for a man who takes a survey of what is already known, not to see what an immensity in every branch of science remains to be discovered."

By way of concrete example, Jefferson settled upon the state of medicine at the time. "The state of medicine," he said,

is worse than that of total ignorance. Could we divest ourselves of everything we suppose we know in it, we should start from a higher ground and with fairer prospects. From Hippocrates to Brown [John Brown, 1735–1788, Scottish physician, author of *Elementa Medicinae* and proponent of his own cult of "Brunonian" medicine] we have had nothing but a succession of hypothetical systems each having its day of vogue, like the fashioned fancies of caps and gowns, and yielding in turn to the next caprice. Yet the human frame, which is to be the subject of suffering and torture under these learned modes, does not change. We have a few medicines, as the bark, opium, mercury, which in a few well-defined diseases are of unquestionable virtue. But the residuary list of the *materia medica*, long as it is, contains but the charlatanries of the art. And of the diseases of doubtful form, physicians have ever had a false knowledge, worse than ignorance. Yet surely the list of unequivocal diseases and remedies is capable of enlargement.

Jefferson then returned to more general interests:

And it is still more certain that in other branches of science, great fields are yet to be explored to which our faculties are equal, and that to an extent of which we cannot fix the limits . . . while the art of printing is left to us, science can never be retrograde; what is once acquired of real knowledge can never be lost. To preserve the freedom of the human mind then and freedom of the press, every spirit should be ready to devote itself to martyrdom; for as long as we may think as we will, and speak as we think, the condition of man will proceed in improvement.

Science and freedom—rationality and liberty—these ideals of the ascendant West were first proclaimed in our land as the living principles of the good society. Through freedom, in Jefferson's lofty vision, men would advance without limit their understanding of nature's laws and of their own natures. Through science, men would liberate themselves from toil and the gross concerns of material want. There had been republics before in the brief history of civilization. They had won the admiration of Niccolò Machiavelli, who advised his Prince: "Republics have a longer life and enjoy better fortune than principalities because they can profit by their greater internal diversity." But these earlier republics had admitted to citizenship only the tiny minority of men who were freed by the labor of others to exercise the powers of government. Our republic was to be the first, born of the mutual fructification of freedom and science, in which all men could be called to the sovereignty of self-governing citizenship.

Today—meaning literally on this happy and solemn occasion—no one can doubt the fulfillment of the eighteenth-century expectations as to science. Certificated and hooded with your doctorates, you have now completed something less than half of the training that is necessary to qualify you for practice or for teaching and research at the present stage in the advance of the science of medicine. Ordinarily, the triumphs of science are measured in the developments that flow from the physical sciences, and these undoubtedly have transformed our environment in those aspects that most powerfully assail our senses. Nonetheless, it can be shown, I think, that the life sciences have brought equal if not greater changes in our lifeways.

Medicine has surely enlarged the list of unequivocal diseases and remedies. While our life expectancy still falls short of Jefferson's own full and vigorous eighty-three years, it is approximately double that of the colonial American. What is more, the relative mortality of men and women has been reversed, signifying the liberation of women from the perils of childbed and a corresponding transformation of the status of women in our society. The falling death rate has been accompanied, until recent years, by a falling birth rate, giving us an entirely new kind of population pyramid—more like a cylinder with equal numbers in each age group, except for the bulge of the baby boom. Hopefully, the deflection in the birth rate will soon be corrected as Americans learn that population control begins at home. Certainly, in any case, the limitation of family size throughout the population constitutes a social revolution of still unreckoned consequence.

In eighteenth-century America, as in the preindustrial or poor nations of the modern world, some three-quarters of the population worked on the land. Today less than 8 percent of our labor force is engaged in agriculture. This is a consequence not only of mechanization but more primarily of biochemistry and genetics—especially in the recent paradoxical contraction of the input of labor and resources into agriculture and the cornucopial outpour from this sector of our economy. Our farmers this year will produce enough food at the farm to feed a billion people. A large percentage of it will be upgraded, via the feed pens, to fat and protein, maintaining the incidence of the strange new nutritional diseases that afflict our population.

The American agricultural surplus may be taken as symbolic of the surpluses that flow from every channel of production in our technology. In the place of scarcity, which has furnished the iron compulsions of the economic and social orders of the past, our economy must reckon with abundance. That we have not yet learned to reckon with abundance successfully supplies a principal theme in what I have to say to you today. At this juncture, we can agree that science has more than fulfilled its role in the eighteenth-century program for the attainment of the good society.

With machines to do the production, the American citizen should by now be liberated for the exercise of his citizenship. As a matter of fact he is: he works fewer hours of the day, fewer days of the year, and fewer years of his life, by very large margins, than his colonial forebear and mostly at tasks that his forebear would not regard as work at all. The practice of self-government should, therefore, be advanced in our land far beyond the standards attained in any republic of the past. With the physical resources now at the disposal of 180 million sovereign citizens, the condition of man in America should be proceeding in a course of assured improvement. If any of these derivations from the American constitutional hypothesis are in doubt—as we must now confess they are—it cannot be through any failure of the term in the equation that is labeled science.

We must look then to freedom. The term "freedom" in the constitutional hypothesis has a special meaning—special in the sense, at least, that it is different from its meaning in contemporary discourse. Freedom is not just an orator's flourish for national interest, prestige, power, or policy. Nor is it a negative word in the sense of "freedom from." Freedom in the social science of the founding fathers had a plus sign. It affirmed the right of the citizen to the fulfillment of his individuality. Stated thus, the notion still requires explanation, for the word "individual," too, has been debased in modern usage—as in "rugged individual," that hero of the war of all against all. Let Jefferson explain; in his letter to the young student, he says: "I am among those who think well of the human character generally. I consider man as formed for society, and endowed by nature with those dispositions which fit him for society. I believe also . . . that his mind is perfectible to a degree of which we cannot as yet form any conception."

Freedom, in sum, is not a state or condition but a process—

the process of individual self-perfection. The citizen is free to the degree that he asserts his freedom.

Nothing in our new knowledge of the dark regions of the unconscious underlying the rational faculty of the mind has compelled any revision in this humane ideal of the eighteenth-century social-psychologists. Nor did they fail to consider the practical implementation of the ideal. Here they looked to political economy: the freedom of the citizen would be secured by his sovereign independence in his material estate. Except for the quarter of the colonial population who were Negro slaves, the typical American was self-employed. By the rational pursuit of his self-interest in the marketplace he would gain the blessings of liberty not only for himself but, through the agency of the "unseen hand," for all his fellow citizens. The facts of life in the infant republic did not, of course, fully satisfy theory. They nonetheless sustain R. H. Tawney's declaration that ". . . the past has shown no more excellent social order than that in which the mass of the people were the masters of the holdings which they ploughed and the tools with which they worked, and could boast . . . 'It is a quietness to a man's mind to live upon his own and to know his heir certain.' "

If freedom resided exclusively in this system of political economy, however, freedom would long since have vanished in this country. Not more than 20 percent of the American labor force is now self-employed. This minority is made up principally of farmers, the small merchants, and the professions. It cannot be said that their pursuit of their self-interest significantly affects the fortunes of our society. In fact, the ranks of the self-employed do not include the true movers and shakers of life in America today. They are employees—members of the giant bureaucracies of the 200 or so largest industrial and financial enterprises that deploy the decisive 50 percent of the assets of our economy. Measured in terms of revenue, the first seven of these organizations come ahead of the largest state government, and the 200 are, on the average, larger than 30 out of the 50 states.

Industrial technology, in all of its interlaced diversity and complexity, has brought a concentration of power and an aggrandizement of power never contemplated in the framing of our Constitution. In the place of the individual citizen, giant institutions now conduct the rational pursuit of self-interest. The unseen hand remains as invisible as before, but the results, on balance, do not show that the public interest is thereby secured and increased. There was general derision at the suggestion that what is good for General Motors is good for the country. There can be no doubt, however, that what is good for any one of these enterprises must be good, if it is not to be bad for the country.

Only a generation ago, the entire system ground to a halt. Long-neglected powers of the federal and state governments were then invoked to establish in America a modicum of the social-security, public-welfare, and resource-conservation measures that had been taken in other advanced industrial societies a full generation earlier. This momentary episode in the evolution of our institutions terminated with the outbreak of World War II. The intervening period of war and cold war has installed a new center of dense concentration at the heart of the still aggrandizing American power system.

The war economy engages directly a full 10 percent of the gross national product and 25 percent of the output of the manufacturing industries. Indirectly as well as directly, it generates 20 percent of the total economic activity of the nation. It derives its forward-driving, expansive impetus at once from military services engaged in doing their duty and extragovernmental enterprises engaged in maximizing their results. Coalesced in common cause, these two elements in the new tripartite system of power in our society have reduced the third element, the civil government, to the status of junior partner.

The working of the system, observed at close range, confronts us with a paradox. At each step in the executive and productive process there is efficiency and rationality. But the sum total of performance is waste and irrationality. The logic of arithmetic pervades the system, yielding correct answers in the steps from A to B, and B to C, but losing inevitably the thread of higher logic that is supposed to relate step D back to A. The decision maker holds responsibility only for those matters within his "decision function"; the military agency looks to its mission, and the contractor to its contract. Considered as a whole, the system is without a mind.

As a result, the nation is burdened with a surplus of weapons it could never bring fully into action even in the event of Apocalypse. Our capacity for overkill has been responsibly estimated at 500 times the destruction of all available targets.

The only nearly sensible justification for this profligate waste of resources is economic: it has served to subsidize business activity and thereby maintain the affluence of our economy. In this function, however, the war economy is failing. Affluence all along has excluded one-third of the nation; unemployment has been rising insidiously with each oscillation in the now well-damped business cycle, and the 1960 census has shown that the middle-income groups are losing families to the sink below the poverty line. Meanwhile, some 15 percent of our installed industrial capacity stands chronically idle.

The public interest, in all its multitudinous and individually insistent forms, has no advocate at the summit of power. The interlocking complexities of the system have long since escaped the deliberative faculties of the Congress. In the presidency there remains a residuum of the power which the office gained in the 1930's, but no President since then has come forward as Tribune of the People. From election to election, under the stage management of impresarios otherwise engaged in the merchandizing of white bread, the two political parties grow increasingly indistinguishable. The daily press and the other channels of mass communication, for their part, hold up a faithful mirror to the mindless system.

Whence decisions come and how they flow through the circuits of power in our society presents as deep a mystery to the individual citizen as the workings of nature. To this it should be added that the mystery remains almost as complete to the bureaucrats, large and small, who man the system.

What I have been describing—with concern and not with dispassion—is the breakdown of self-government. The power of the sovereign citizen has been pre-empted to institutions that owe no legal accountability to the public interest or, under the mask of secrecy, render none. While I do not think that this illness is confined exclusively to our body politic, I have reviewed the American picture in detail because it was our country that first held out the promise of self-government. If there is any quarrel with my presentation of the diagnosis as to our own land, I am sure it will be accepted, with appropriate amendment and amplification, as a valid description of the power system arrayed against us in the cold war. In fact and in essence, the same analysis applies to the relation of citizens to power in all of the modern nations, with variation over the range between these two extremes. The promise of self-government, held out with such generous and humane vision less than 200 years ago, has given way to decerebrated government by spinal reflex.

The issues of our age go undebated in the open in the public forums charged with this central function of self-government. There are only two issues worth talking about. The first, of course, is War: whether any value or purpose has yet been discovered in the social and political evolution of our species that can be served by the prospective incineration of perhaps a third of the world population and all of the seats of civilization. The second is Peace: the mobilization of the material and intellectual resources of mankind to make actual the now technologically feasible abolition of poverty in the world. The two issues are, of course, one.

Meanwhile, who can doubt, as current events oscillate from drift to crisis, that we are living in a prewar period? We cannot hope for an indefinite prolongation of the present twilight. Mindless power systems are without heart as well, and they know neither fear nor mercy.

How can humanity be restored to power? By what means can the citizen reassert his sovereignty? In attempting the difficult task of framing an answer, I have consulted the record of two actual undertakings toward this objective. Together they suggest the range of possibilities in which each of us may find his own course of action.

Birmingham, Alabama, during the past eight weeks has won a new place in our history as the scene of the latest episode in a nationwide campaign by the American Negro to win his citizenship. With remarkable discipline, arising from an exquisite sense of the moral question at issue, Negroes of all ages and from all walks of life have pioneered new modes of action hitherto unknown to the American way. From the courtroom to the lunch counter, by word and by deed, they have been teaching their fellow citizens that they are equals in fact as well as in law. In countless instances of nobility and courage they cannot fail to have inspired among the rest of us a new faith in the dignity of man. They have shown how to assert that dignity on other issues felt as deeply.

In quite another realm of issue and action, I cite the example of the small group of scientists from nations on both sides of the cold war who have managed to meet periodically with one another over the last seven years—not as representatives, nor even as citizens of this or that country, but as men—to talk rationally and honestly about war and peace. In all of their various gatherings the total number of participants in these "Pugwash" meetings has not exceeded 300, and the regular participants in these conferences have numbered no more than two dozen. For some of these men, on both sides, it has required great courage even to attend these meetings and, in the presence of their respective countrymen, to follow the logic of informed discussion to new conclusions. From their meetings have come understandings and agreements which the participants from both sides have carried home and laid before their governments in private and their fellow citizens in public. Certain of these agreements have had historic consequence. They provided the technical foundations for the first efforts to negotiate a ban on the testing of nuclear weapons; in the more recent "black-box" scheme they have provided a basis for renewed effort to negotiate a test ban. Beyond the preliminary step of a test ban, they have supplied the intellectual leadership for the design of an ultimate disarmament agreement. As each of these men can testify, it has proved far more difficult to sell these agreements to inert bureaucrats and insecure politicians at home than to arrive at these agreements in earnest debate with colleagues from the other side of the cold war.

Between the two ends of the spectrum of action thus opened up by courageous and honorable men and women, there are examples for the consideration of all citizens. The obligation to explore these possibilities falls with particularly heavy weight upon those who are engaged in science. Their training and talent provide the primary source of power to the power systems of the world. They occupy strategic positions from which to take effective action—whether it be in the choice of their life-work, in setting the terms of their work, in seeking the truth, or in speaking it.

Within the scientific community the obligations of citizen-ship must press with increasing intensity in the future upon workers in the life sciences. The principal present dilemmas of society come from the relatively simplistic world of physics. So much the less can the existing power systems be entrusted with the knowledge now issuing from molecular biology and from exploration of the reticular system deep in the fissures of the brain. Knowledge necessarily implies control. Under any government but self-government the control of human heredity and the power to manipulate human behavior must inevitably become instruments of tyranny beyond the most cruel precedents of history.

In the last paragraph of his letter, Jefferson said to his young correspondent—and to us:

The generation which is going off the stage has deserved well of mankind for the struggles it has made and for having arrested that course of despotism which had overwhelmed the world for thousands and thousands of years. If there seems to be danger that the ground they have gained will be lost again, that danger comes from the generation your contemporary. But that the enthusiasm which characterizes youth should lift its parricide hands against freedom and science would be such a monstrous phenomenon as I cannot place among possible things in this age and this country.

As physics, over the past century, has led philosophy to questions at the foundations of knowledge, so biology now confronts us with ultimate questions of human value. Plainly, the task of the physician must be the healing of society.

The Sorcerer's Apprentice

THE American Federation of Information Processing Societies brings together the creators and users of the technology that now carries the thrust of the industrial revolution into every corner of our economy. This essay was presented at the AFIPS Fall Joint Computer Conference in San Francisco in October, 1964.

THE computer is the engine of this latest phase in the acceleration of the industrial revolution. The role of the computer cannot be measured in the simple terms of the number of computers at work in the American economy or even in the extraordinary variety of functions in which the computer has found work to do—from accounting routines to industrial process control to creative enterprise in mathematics itself. More significantly, computer technology gathers in and brings to intense focus the most diverse discoveries on the frontiers of knowledge—from investigation into the nature of matter to speculations at the foundation of knowledge. It is the agency through which the advance of human understanding now finds its way to the control of natural forces in time intervals that grow shorter year by year and month by month.

Because the time lag between invention and application now diminishes so swiftly, it becomes possible—and necessary—to forecast the ethical, social, and economic implications of this development. Today in our country and in certain other industrial nations, men are compelled to recognize and give assent to a profound transformation in human values. Technological change has already largely eliminated people from production; it has sundered the hitherto socially essential connection of work to consumption. The citizens and the institutions of these nations must accommodate themselves to the law of material abundance: each individual can secure increase in his own well-being only through action that secures increase in the well-being of others.

This novel dispensation stands in contrast to the law of scarcity which, in the words of that subversive nineteenth-century Russian patrician Alexander Herzen, declares: "Slavery is the first step toward civilization. In order to develop, it is necessary that things should be much better for some and much worse for others; then those who are better off can develop at the expense of the others."

The iron law of scarcity underlies the ethical dilemma of political economy which has sought for nearly three centuries to discover or to rationalize equity in social institutions long ago designed to secure the inequitable distribution of goods in scarce supply. Adam Smith, the first great systematizer of economic theory, was foremost a moral philosopher. In his *Theory of Moral Sentiments*, published in 1759 and the work which brought him his principal contemporary fame, he traced the roots of moral action to the "passion of sympathy"—"which leads us into the situations of other men and to partake with them in the passions which those situations have a tendency to excite." It was later, in the *Wealth of Nations* published in 1776, that he undertook to explore "those political regulations which are founded, not upon the principles of justice, but that of expediency, and which are calculated to increase the riches, the power and the property of the state." Against the princely mercantilism of the autocratic Continental powers, Smith asserted the labor theory of value: "Labour is the real measure of the exchangeable value of all commodities. . . . Equal quantities of labour at all times and places are of equal value to the labourer. . . . Labour alone, therefore, never varying in its own value, is alone the ultimate and real standard by which the value of all commodities can at all times and places be estimated and compared." In the free play of supply and demand in the open market, the products of human labor were to find the just and equitable price at which they were to be exchanged. In the market, labor, itself a commodity in consequence of the division of labor, would also find its fair price. Under the sure guidance of the "invisible hand" each man could seek his private interest, confident in the knowledge that he thereby secured the public weal.

For the generations that launched the industrial revolution in eighteenth-century England, Adam Smith and his successors in political economy furnished not only the guidelines to practical action but the moral assurance necessary to the taking of action. Before the middle of the nineteenth century, however, it had become impossible to conceal—in the blight laid upon green England by the carboniferous phase of industrialization—the failure of their enterprise. Benjamin Jowett, Master of Balliol and translator of Plato, spoke for the alienation of the humanities from the sciences when he said: "I have always felt a certain horror of political economists since I heard one of them say that the famine in Ireland would not kill more than a million people, and that would scarcely be enough to do much good."

Even as Jowett wrote, the first phase of the industrial revolution had made such computations obsolete as well as patently immoral. In 1863, the year of the Emancipation Proclamation, mechanical horsepower generated by steam engines in the U.S. economy exceeded for the first time the output of biological horsepower by horses and men. As early as 1900, only 75 percent of the U.S. labor force was employed as "producers of goods"; more than half of these producers were engaged in farming and the next largest percentage in unskilled labor functions. By 1960, human muscle had been all but eliminated from the production process. The census for that year shows that less than half (45 percent) of the labor force was now employed as producers of goods; farmers (7 percent) and un- skilled laborers (5 percent) were approaching statistically negli- gible percentages of the labor force. More than half of the producers were classified as "operatives," that is, human ner- vous systems still interposed in process control feedback loops not yet completely closed by electronics.

In the present phase of acceleration, as is well known, the industrial revolution is eliminating nervous systems from the production process. Robots—artificial sensory organs and mechanical controllers linked by feedback circuits—have been taking over from human workers in all of the fluid process industries. In at least 85 plants in the United States, computers at the center of control networks have transformed the process streams into truly self-regulating systems. The computer and the feedback control loops have now begun a corresponding transformation of the discontinuous processes of the metal- working industry. The same revolution in technology—for ex- ample, transcontinental pipeline transportation of fluid com- modities under computer and feedback-loop controls—is under way or impends in all of the production sectors of the economy.

During the past decade, blue-collar employment in American manufacturing has actually declined, while the output of these industries has nearly doubled. The rate of increase in productiv- ity in the production sectors of the economy, which has aver- aged 5.6 percent over the decade and has been accelerating, is grossly understated by productivity figures applied to the entire labor force. These, the figures given widest circulation, have shown an annual improvement of only about 2.5 to 3 percent.

Until recently, increase in employment in trade and distribu- tion and in the services has compensated for disemployment from production. The computer, however, finds application even more readily in the functions that employ human beings in these sectors. The "white-collar" computer, equipped with a keyboard on its input and output side, is conceptually a much simpler organism than the computer equipped with sensory organs and muscles that displaces the blue-collar worker. A conservatively estimated millionfold increase in the data-processing capacity of organizations equipped with computers as compared to organizations manned by human beings and assigned to comparable tasks has already been demonstrated in military command and control systems. Although computer technology has barely begun to find its way into trade and distribution and the services, increase in employment in these sectors has already begun to slacken. In the private sector of the economy it now barely offsets disemployment from the production industries. During the five-year period from 1957 through 1962, the private enterprise economy generated less than 300,000 additional new jobs.

The creation of new jobs in the economy as a whole has now lagged the growth of the labor force for more than a decade. This is a polite way of saying that the economy is afflicted with a constant and insidious increase in unemployment. Ever since 1952, the rate of unemployment has been larger at the peak of each ripple or boomlet in the economic cycle, and each recession has left a larger percentage of the labor force high on the beach.

Debate continues as to whether the country's rising unemployment is "cyclical" or "structural." Classical economists—and nowadays Keynesian economists are "classical"—assure us that the unemployment is cyclical. They point to the history of the past sixty years in stubborn support of the thesis that the labor-saving effect of technological progress merely frees labor from one task for employment in another. It is conceded that frictions make for unemployment in this turnover of the labor force, especially when progress goes forward rapidly. But sooner or later new jobs, generated by ever greater economic activity and an ever expanding gross national product, soak up the unemployed. By tried and tested and now generally sanctioned countercyclical measures—for example, by the recent federal tax cut—the fluctuations of the system can be damped and the peaks and valleys of unemployment smoothed out. When the

Kennedy administration took office, its official economists were arguing that unemployment at the rate of 4 percent could be regarded as normal. Despite the tax cut and the prolongation of the present boom, unemployment now ranges above 5 percent.

Increase in unemployment accompanying expansion of economic activity would seem to indicate that a rising percentage of the unemployment is indeed structural—that people, in other words, are being displaced from the economic system in ever larger numbers by mechanization, more specifically by the computer and its accessory and allied technologies. Consider, for example, the computer industry itself, thus broadly defined. If employment were to expand in any industry during this period of intensive mechanization, one would think first of the payroll of the industry that is doing the mechanizing. What is more, the technology of electronics that furnishes your hardware has been notably, if paradoxically, highly labor-intensive. Until a few years ago, labor would represent up to 60 percent of the production cost of a piece of electronic hardware. Engineering would constitute the major investment; materials would be a minor cost and capital equipment a negligible item on the balance sheet. In these respects electronics was like the garment industry: a business anybody could get into, providing he had a bright idea and could finance his payroll long enough to get his product on the market. Within the last ten years, as I need not tell you, electronics has gone solid state. The transistor and the micromodule are even now yielding to the integrated circuit. With this development, acre after acre of workbenches at which housewives and high school girls wield pliers and soldering irons has been disappearing. Labor cost is vanishing in the economics of electronics. Material costs have now become significant; engineering and plant costs, transcendent.

In other words, the prevailing relationships among the factors of production in electronics are being turned 180 degrees around. With people being exiled from the computer industry as rapidly as the computer itself is promoting the disengagement of people from jobs in other sectors of the economy, the expansion of this industry will not generate anything approaching a corresponding build-up of its payroll.

It cannot be said, any longer, that the industrial revolution is the same old story. The acceleration of technological change, driven by the accelerating advance of human understanding, reaches to the very heart of the institutions of our society; that is, to the value system upon which those institutions rest.

The unemployment figures present a profoundly misleading measure of the degree to which our capacity to produce material abundance has outrun the capacity of our institutions to secure the distribution of that abundance. It must be reckoned, in the first place, that some 8 million persons are employed in the war economy and contribute nothing whatever to the flow of material abundance from our nonmilitary productive system. If the production workers in the war economy are subtracted from the productive work force, then the percentage employed as producers of goods falls below 40 percent. But this figure still overstates the truth because most of the goods circulating in commerce and consumed by American citizens are produced by the very much smaller percentage of the labor force that is employed by our most efficient production organizations.

Consider, for example, our farms. Some 85 percent of the food that moves from the farms to the markets comes from less than one million farms. The same is true of industry; the few large and efficient corporations in each industry, with their relatively smaller payrolls, produce the overwhelming percentage of our industrial output. If a small minority of our working force is today doing most of the production, then, in the future, we can expect to see an even smaller minority of our working force account for all of the production of goods in our economy. The sorcerer's apprentice has thrown the switch. The great test of our democracy is to find ways to distribute or dispose of the mounting flood of abundance.

Actually, by the kind of improvisations that are so characteristic of democracy, we have had some success in coping with this task starting from the turn of the century. In 1900, 40 percent of the adults of our country were not employed; that is to say, they were either unemployed or they were not in the labor force. In those good old days, 57 percent of the adults of the country were employed in the private sector of the economy. Our country still approximated the description it gave of itself in the Declaration of Independence, as a people engaged in the pursuit of happiness—in the pursuit of private interest, either their own or that of their employers. Only 3 percent of the American people were on the public payroll. In 1960, the same 40 percent of our population was not employed, either unemployed or not in the labor force. But only 40 percent of the population was now employed in the private sector of the economy. A full 20 percent of the American people found their employment either directly on the public payroll or indirectly through the increasingly huge expenditures of governmental agencies for the product and services of private corporations— not only in the war economy but in the construction of highways and other major public works ventures. In round numbers, half of all public revenues return to private incomes through governmental purchases from private enterprise. Thus it is that one-third of the working force now owes its employment to public expenditure.

Direct employment in the public sector has been increasing at five times the rate of increase in the private sector. During the past five years the public sector generated more than a million of the less than 1.5 million new jobs in the economy. Since the federal payroll remained constant during this period, this gain must be credited to state and local governments. It can be declared with pride, furthermore, that the biggest part of the gain was in the payroll of our public education system. This, in turn, may be taken as an indication of the responsiveness of our value system to the evolutionary pressure of abundance.

The national accounts also indicate, however, that the evolution of our social institutions is falling behind the accelerating pace of technological change. It turns out that the magnificent industrial apparatus of America has been producing as much poverty as wealth. Poverty is now officially acknowledged to be the lot of at least 25 percent of our population. Contemporary American poverty is selective, as Michael Harrington has pointed out. It tends to settle in places where it disappears from sight—hidden away geographically, for example, in Appalachia and in the central cities from which more fortunate members of our society have fled to set up their new settlements in the suburbs. In New York and Chicago, the third generation of families on relief has already begun its blighted existence.

Poverty is selective also with respect to age. Unemployment rates, which for the labor force as a whole are officially acknowledged to exceed 5 percent, exceed that rate among the youth by at least twice, and among Negro youths the rate exceeds that among white youths by more than twice again. In fact, the prevailing rate of unemployment among Negro youths in the slums of our central cities runs from 40 to 50 percent. The high school dropout may spend five years or more in empty limbo between school and his first job. Out of such alienation of so many of our young people has come the rise in juvenile delinquency, and out of the rejection of our Negro youth came the riots in the cities of the North during the past summer. Poverty is equally selective with respect to age at the other end of life. The 40 percent of our adult population not counted in the labor force now includes several million men and women retired to live on the pittance of monthly social security checks, under contract not to seek gainful employment.

Such are the shameful facts that confront us in the midst of the most prolonged boom since the crash of 1929. Forecasters predict the boom will hold up well into the first quarter of the new year. Against the expectations of myself as well as a few other pessimists, the tax cut has had a strongly stimulating effect on the economic system. It has encouraged a remarkably high rate of investment by industry in new capital equipment— one-third of the investment going to modernization, thereby also accelerating the rate of mechanization. Through the action of the familiar Keynesian multiplier, these expenditures on the capital investment side have helped to sustain consumer expenditures at new highs. The argument that fiscal measures may help to reduce unemployment, therefore, finds support in the current movement of the economy. Although these measures and the prolongation of the boom have not actually reduced unemployment below the 5 percent line, a catastrophic increase in unemployment has been forestalled.

The financial pages all agree, however, that this boom has a terminal date; most set it around the end of the first half of 1965. As the boom runs out, the application of mere countercyclical measures—a further cut in federal taxes, for example—will be of no avail. At the same time, responsible citizens and public officials must face up to the question of the armaments budget. Even in advance of that distant date when we may see some substantial measure of disarmament, the military budget must be cut back. Our country long since acquired the capability of overkill, counting all the targets in China as well as in the U.S.S.R. Yet, with the business cycle turning downward, it will take brave men to cope with the fact that 8 million jobs hang directly upon the size of the military budget.

Plainly, the termination of the present boom will require not a tax cut but, on the contrary, a considerable expansion in public expenditure. That expansion has got to come, moreover, in the federal budget. It is perfectly plain that the payrolls of local governments are not equal to short-run challenges; they cannot respond as flexibly and with the same massive effect as federal expenditures can. The next administration will be compelled to seek, therefore, a vigorous expansion in federal expenditures on public works and public welfare.

I don't think I betray any secrets of the present administration at this point in the national election campaign by telling you that task forces in every department in Washington are at work on the question of how to spend increased sums on nonmilitary undertakings of the federal government. The house economists of the Kennedy administration observed some time ago that the nation had accumulated a backlog of demand for public works and welfare equal in magnitude to the backlog of demand for consumer goods and capital goods at the end of World War II. The Arms Control and Disarmament Agency, which is principally responsible for analyzing the prospective impact of disarmament on the economy, predicts an easy transition from huge outlays for warfare to huge outlays for welfare— it points to this backlog of unmet public needs. Soviet economists join their American colleagues in assuring us that capitalism is equal to the task.

All of this is cheering to hear. And it is especially considerate of the Soviet economists to give us their encouragement. But, against a value system that stoutly resists every increase in federal expenditure except those that carry the absolute sanction of the national defense, any effort to increase public expenditures for public welfare will encounter heavy political opposition.

The backward state of our value system is suggested by the following figures describing the condition of our society: America has, in fact, the highest rate of unemployment among all the industrial nations of the world. If the maintenance of adequate nutrition is taken to establish the poverty line, then Department of Agriculture studies show that not one-quarter but one-third of our fellow citizens remain ill-fed. Our country has the lowest ratio of public to private expenditures, even with our gigantic war budget. In the public sector—in federal, state, and local budgets—our economy turns over 25 percent of its gross national product. The lowest figure you find in any other industrial society is 30 percent, and those budgets are nowhere so heavily discounted by military expenditures as in our country. America has the lowest rate of public expenditure on public welfare and public works; it comes to something less than 10 percent for the country as a whole. The lowest figure in any other industrial nation is nearly three times this percentage.

Last spring, the Johnson administration took its first tentative steps to meet the impending short-run economic crisis. It assembled from already ongoing and funded activities of the federal government an antipoverty program. Meanwhile private institutions and individuals were attempting to draw the lines of long-run perspective. One committee of concerned citizens—the self-styled Ad Hoc Committee on the Triple Revolution, which included political economists, historians, former public officials, labor leaders, civil rights workers, and at least two men who have met payrolls—looked rather more deeply into the widening gap between the productive capacity of our industrial system and the effective demand of our consumer economy. In one conclusion to their analysis, they declared: "Society, through its appropriate legal and governmental institutions, must undertake an unqualified commitment to provide every individual and every family with an adequate income as a matter of right."

The idea of paying people incomes whether they work or not captured attention in newspaper city rooms all across the country. It seems scarcely necessary to add that the idea also won a great deal of unfavorable comment. Setting aside the ephemeral essays of the commentators and pundits who explain the news to the rest of us, the comments of two distinguished public figures are illuminating. The Secretary of Labor, Willard Wirtz, declared: "I think the analysis is right but the prognosis and the prescription is wrong." He added: "I don't believe the world owes me a living and I don't believe it owes anyone else a living."

The other comment comes from a man who was, at the time, candidate for the Republican presidential nomination. You may recognize his voice. He said: "Our job as Republicans is to get rid of people who will even listen to people who say we should pay people whether they work or not!"

These two statements, taken together, speak faithfully for the austere premises of classical political economy and the tenacity of their grip upon the conscience of many members of our society. Yet the preposterous notion of a guaranteed annual income (or GAI, as it has come to be called) has found its way onto the agenda of public issues.

Upon deeper reflection over the summer, for example, *Life* magazine returned to the subject for the second time on its editorial page. This time, *Life* conceded that there is technological disemployment: ". . . experts can't agree whether technological unemployment is growing by 4,000 or 40,000 jobs a week. But it is growing fast enough to see that the seeming logic of the . . . plan for free incomes, or instant socialism, may grow too."

Having frightened itself with this prospect, *Life* goes on to say that there can be "more than one radical alternative" and puts forward one of its own: "It is private capitalism, after all, that has brought us to the brink of this daunting affluence, and there is an obvious capitalist solution to the problem that the success of capitalism is creating. It lies in the ownership of the machines and the processes that are destroying the old jobs and creating the new wealth." *Life*'s proposal is that the ownership be spread—to everybody! Endorsing the analysis set forth in *The New Capitalists* by Mortimer Adler and Louis Kelso, *Life* would ". . . guarantee bank loans for new stock acquisitions through a Capital Diffusion Insurance Corporation modeled on FHA."

Let us tarry a moment, here, to contemplate the prospects of instant capitalism. The figures indicate that it would be much more difficult to achieve *Life*'s worthy purpose by instant capitalism than by what it calls instant socialism. Consider these disparities in the wealth of our citizenry: as is well known, the bottom 20 percent of our population gets only 5 percent of the national income—at the summit of society these percentages are precisely reversed. The bottom 20 percent thus does poorly enough as income earners. But they show up even worse as capitalists: they hold no liquid assets whatever, except the cash they may happen to have in their jeans. The next 30 percent of the population above holds liquid assets not exceeding $500 per family. So the bottom 50 percent of our society holds less than 3 percent of our liquid assets. It scarcely need be added that these people have no equity or debt interest in the productive system of our land, for 87 percent of the people have no such stake in the system. As for real property, 50 percent of our population have an equity of less than $1,000 in the homes in which they live. You have to go to the top 10 percent of income earners before you find people whose assets begin to equal their annual income; you have to go to the top 1 percent before you find people whose equity and property holdings keep them in the style to which they are accustomed. Plainly the proposal that we seek a more equitable distribution of affluence through the redistribution of ownership presents a more radical alternative than the achievement of that end by the redistribution of incomes.

Life is not alone in its concern with the question of how American society might now go about the equitable distribution of the abundance that overburdens institutions framed for the opposite purpose. That soft-spoken quarterly *The American Scholar*, the journal of Phi Beta Kappa, devotes most of this quarter's pages to a symposium on "the problems that unite us." Out of six articles in this symposium, four plainly reflect thoughtful consideration of the possibility of guaranteeing incomes to people. I will quote from one author, August Heckscher, a perceptive and sensitive student of American life who served the Kennedy administration as the President's special assistant in cultural affairs. He begins by saying: "The objections to this approach [that is, the guaranteed annual income] are obvious," and declares: "The very idea of large populations doing nothing but pleasing themselves goes against the American grain." Nonetheless, he comes forward with a radical alternative of his own: "Suppose the monetary rewards of society went, as now, to those who work—and also to those who study. Would this not be a means of assuring their being saved from a bored and barren existence?"

This author then goes on to suggest other ways in which the surplus of human capacity might be soaked up: "At the simplest level one can readily conceive efforts to organize personal or household services more imaginatively so that the work can be done more efficiently. Hours can be made regular and wages can be more nearly commensurate with those earned in other fields." A little later in his analysis, touching on the question of how these increased wages to domestics are to be financed, he comes up with a truly radical alternative: "The salaries . . . could be supplemented [from the public treasury] so as to keep the supply adequate and yet not put the wage out of reach of those who require such services. To supplement in a similar way the rate which people are ready to pay handymen or gardeners could substantially cut relief rolls."

This surely goes beyond either instant socialism or instant capitalism; you might call it instant feudalism. In fact the vision of instant feudalism comes clearer in this author's next, still "more far-reaching" suggestion: "It assumes retirement from the industrial work force at a considerably earlier age than now, together with pensions and social security which would be clearly conceived as 'deferred wages.' . . . From such a pool we could draw a host of talents and services which would make our common life more various, colorful, and pleasant. . . . We can indeed conceive a whole second economy—the economy of craftsmanship and service—growing up alongside the economy of the machine."

Probably, this vision could be more swiftly and effectively realized in certain of the underdeveloped countries where the economy of craftsmanship still exists and where it is threatened by destruction through the infectious spread of the industrial revolution. In America we would have to reconstruct the economy of craftsmanship from the ground up.

Before we start designing Utopias or building the Great Society, it seems to me, we ought to turn to a more searching and possibly painful re-examination of our inexplicit premises—our values. A good way to begin is to ask what we mean by work and what we mean by leisure. With these two words we precipitate the ethical crisis. The proposal of a guaranteed annual income presses the underlying issue in its sharpest and most uncomfortable form.

The objection to the Heckscher vision of the dual craftsmanship-machine society rests upon its hierarchical character, implicit in the compulsion that relates the services of the handyman and the gardener to "us." This defect could, in fact, be cured by the guarantee of an annual income, paid as a matter of right and not in compensation for services rendered. There would then be no reason why the culture of craftsmanship and machines could not flourish side by side in moral parity. And there could even be a third culture—of leisure, which would include, I hope, dry-fly fishing.

On the other hand, criticism of the GAI notion from the left expresses the dark suspicion that this is a middle-class stratagem to tranquilize the proletariat by putting the poor on the dole. Apparently, as Gerald Wendt of UNESCO has observed, most people are deeply troubled by the thought of what *other* people might or might not do with their leisure time.

Except for the attention it has so recently won in public discussion, there is nothing very novel or profound about the idea of a guaranteed annual income. Nor is it so novel in practice. A substantial portion of our society is already living not on a guaranteed and not on a securely annual income but on an income from the public treasury. The people get these incomes on the most humiliating and degrading terms. They get their dole because they present themselves for certification by the appropriate authorities as indigents or paupers; or they get their monthly checks from Uncle Sam because they take an oath not to go back to work and earn more than a stated percentage of their social security income. In other words, the American society today offers an income without work to a large number of its members but makes the offer on terms that shame us all. The ugly transactions involved derive their ethical justification from the deep unconscious of society—from the institutional memory of the days when the slave-master drove 80 percent of the population to work in the fields and mines in order that the few might get on with the high occupations of making history and civilization. The cruelty and inhumanity that persist in our system from those days must be extirpated if we are to resolve successfully the issues that confront us in the tide of abundance set running in America by the present culmination of the industrial revolution.

In my opinion, the issues must be met under two major headings. First, we must recognize that economic and social institutions are man-made and so subject to human will. We can't see the invisible hand because, in truth, it isn't there! The enormous power conferred upon modern societies by industrial technology must be brought under the witting and rational control of democratic institutions still to be perfected.

Second, we must recognize that abundance sets the foundations of an entirely new ethical and moral order. The cultural deprivation that blights the life of a single child in the racial ghettos of our central cities ultimately exacts its cost in the lives of every other citizen. The prolongation of the agony of economic development threatens to destroy the frail parliamentary institutions of India and bring that poverty-stricken nation into the nuclear club under a military dictatorship.

At this turn in human affairs it is plain that each man's well-being can increase only to the degree that the well-being of all other men is increased. The work of the world still remains in large part to be done. But the instruments to accommplish it are now in our hands. The work that needs most to be done, especially here in America, is in tasks that enrich society as generously as the individuals who undertake them.

IV The Right to Health

Metropolitan
Medicine

AMERICAN medicine presents not only a scale model of the American system but also something of a caricature. It is a technology that draws upon deep new understanding of the most intricate process we know of in nature. It is a small business that stoutly defends the classical ideology and still approximates more closely than any other the model that pits numerous producers in competition in the open market. It is a big business that secures generous public subsidy with minimum of political constraint.

Along with the larger system, the medical economy is now in crisis because its operations have drifted so far out of relation to public need. Few families get the care they ought to have, the early and often preventive care now called "health maintenance." None can really afford the care they are compelled to seek when overtaken by accident, illness, and age.

The future of American medicine is without doubt prefigured in the evolution of the local medical economy of New York City. Here more than 40 percent of the practitioners are employed full time on salary by institutions. In a dozen of these institutions are concentrated the major intellectual and material resources of the system. The public treasury supplies nearly

50 percent of total revenues, steeply increased from the 30 percent figure for 1962, cited in the second essay in this section. Third-party payments account for a major portion of the balance. The once jealously defended boundary between the public and the "voluntary" sectors has been blurred further by the devolving of the municipal hospitals out of the city into the custody of a public benefit corporation charged with devolving these hospitals still further into control and management by community trusteeships. The city health agencies, brought together in a single administration, are learning to function as ombudsman and agent for the public in contracting for services to be rendered at public expense. The public in turn, through increasingly effective "health action" groups, is finding ways to articulate needs not automatically asserted by public and third-party payments. New York medicine is emerging as a new kind of socioeconomic entity, a "social utility," in the phrase of Eveline Burns.

A strategically decisive feature of this social utility has been the Health Research Council of New York City. Just at the time, in the late 1950's, when federal agencies were taking over the subsidy of science from all other benefactors, a group of local medical scientists and public health officials persuaded Mayor Robert F. Wagner to establish the city's own granting agency. The small sums deployed by the Health Research Council encouraged work on an order of priority different from that financed by federal money, so often with motivation supplied by hypochondria in Congress. I had occasion to explain the special features of this civic invention to a meeting of its career scientists in December, 1965, as set out in the essay that follows.

To my shame and sorrow as a New Yorker, I must now record the impending demise of the Health Research Council. The inflation in the city's "uncontrollable" expenditures—the bills for current welfare, education, health, public-safety, transportation, etc., services—left no funds over in fiscal 1971–1972 for discretionary investments in the future, not even for investments that promise some measure of control over those uncontrollable expenditures. This, at least, was the judgment of John

V. Lindsay and his colleagues at City Hall. They put the Health Research Council down in the budget, accordingly, for a terminal grant. An "unbreakable" contract, written to the specifications of Fiorello La Guardia, kept them from doing the same thing to another civic invention unique to New York City: the distinguished Public Health Research Institute, which had been established on City funds in 1941.

In this debacle, I have to confess a personal failure. On appointment by the mayor, I had chaired a committee to evaluate the Health Research Council; we had concluded that this was not only a legitimate municipal activity—for many of the same reasons as are set out in the essay that follows—but one that had already contributed to reduction in the city's current uncontrollable expenditures in health services. This effort in the popularization of science at close range persuaded neither the mayor nor any of his counselors—"generalists" all, but with a blind spot for science in their vision, and made blinder by the anxiety that illiteracy in science so often excites in the most educated. I find what comfort I can in knowing that Fiorello La Guardia and Robert F. Wagner would have fought to save both institutions. But, then, neither of them would have needed any counsel from me.

THE Health Research Council of the City of New York is the vehicle for a unique experiment in the making of public policy for science. This agency spends city money for the support of fundamental research. On the principle that the man is more important than the project, it makes its biggest outlays in career grants to individual scientists. These grants are now paying the stipends of 172 biomedical scientists; all that is asked of them is that they do work that they think is interesting and important—and that they do it here in New York City.

When I called this enterprise "unique" in my first sentence, it was with civic pride. I now repeat the word with some anxiety. The Health Research Council has not been replicated by any other city, even though the example has been before them now for nearly six years. That reminds us that we have a difficult question to answer this year and every year we manage to keep this experiment going. Science knows no national boundaries, and so much the less does it recognize the city limits. The same must be said of the ills afflicting mortal human flesh. Then should the city support science? The $8 million that the Health Research Council spends from the city's hard-pressed revenues would show up just this side of the decimal point in the hundreds of millions that are spent by the National Institutes of Health in the support of fundamental research in the health sciences and general biology. Why should Father Knickerbocker add this pittance to the "gigabucks" that Uncle Sam lays out for fundamental research?

The Council has faced this question from year to year as it has steadily increased its commitment to its career scientist program. If we have no answer, we at least have an argument. First of all, by bringing 172 first-rate biomedical scientists to New York City and keeping them here, we have measurably strengthened the six medical schools in this city. For the purpose of persuading the Board of Estimate and the City Council to renew our appropriation from year to year, the argument must be carried another step further: the strengthening of the medical schools secures improvement in the medical care that is administered in the city hospitals.

At the beginning of the Wagner administration, the condition of the New York City municipal hospitals had become one of the scandals of this country's medical establishment. The affiliation contracts that now tie each of the municipal hospitals to a medical school or strong voluntary hospital have brought high-quality medical and surgical services to the city's medical dependents. Some career scientists have their laboratories in the municipal hospitals. More important, their presence on the faculties, wherever they work, has fortified the medical schools to take on this vital assignment. With no more than one string tied to the Career Scientist Awards, therefore, we are confident that we have already got the taxpayer his money's worth.

There is still another covert strategy in this experiment in municipal science policy. With this strategy working, the Health Research Council expects to see a speculative return on the city's investment in the Career Scientist Awards. We expect that those scientists, just because they are working here in this city, are going to get interested in the health problems of its inhabitants. Because they were chosen for excellence, regardless of project, we expect they are going to do excellent work on the city's problems. As the yield comes in from this development, we hope that the city fathers will not mind discovering that funds from the city's tax base have gone to help the denizens of other cities as well as ours.

New York City is the model, perhaps unhappily, of the habitat of the entire human species in the near, predictable future. Already 70 percent of the population of the United States lives in urban communities; some 20 to 25 percent of the population of the country lives in the three great conurbations of the East Coast, the West Coast, and the Great Lakes. Within another generation more than half of our people will live in a predicted six "conglomurbations" that will be located on the East Coast, the West Coast, the Great Lakes, the Southwest, the Southeast, and the Mid-South. Any medical research that is good for the people of New York will necessarily be good for the people of the United States. And, it may be added, of the world; the same titanic process of urbanization is under way everywhere on our planet. At the turn of this century there were only 30, or so, cities in the world with a population of a million or more. Of those cities, nearly 20 were in what we now call the developed countries. Today, there are 65 cities of more than a million population, some half-dozen with populations exceeding 10 million. Of these cities, more than half are in the underdeveloped countries. These cities of the poor have arisen without the benefit of the technology that makes life possible for 8 million people within the 400-square-mile territory of the city of New York.

What are the health problems of people who live in the city, this new habitation of mankind? The frightening fact is that nobody knows. Statistics as to the morbidity and mortality of our population are heavily qualified by the realization that the profession of medicine tends to diagnose and record only those conditions that it recognizes and can do something about. Lawrence J. Henderson, who moved over from the Harvard Medical School to the left bank of the Charles thirty-five or forty years ago, said:

. . . Upon the appearance of the bacteriological period in the history of pathology, an intellectual decline, as I think it may be fairly regarded, set in. In accordance with the simple views of Pasteur, the specific virus came to be regarded as the cause of each infectious disease and this view could hardly be questioned during the time of rapid progress in discovery of microorganisms. The hypothesis is roughly correct, but it has obscured the organic character of disease and the mutual dependence of the many variables which must be taken into account, if the state of the patient is to be understood.

What the science of metropolitan medicine requires is the invention of a new high-style epidemiology. The instruments must be tuned not only to sort out infectious organisms but to recognize and measure all the other factors—social, economic, and ecological—that go to the definition of health and disease.

The Health Research Council, in addressing itself to this unknown territory of metropolitan medicine, has so far identified three major afflictions. The first is drug addiction. If you would learn what role this plays in the social pathology of this city, just spend a morning in Felony Court in any of the five boroughs. The working party of the Health Research Council organized by Vincent Dole to find a new approach to drug addiction has already some impressive leads. The power of social and psychological forces, along with that of heroin, is suggested by one finding. Because the junk on the market in the city is of such uncertain adulteration and dilution, the addict often fails to get a physiologically effective "fix"; yet he will enjoy his "high," all the same, followed by the agony of his "low." At present, Dole and his colleagues are developing confidence that a narcotic called methadone may substitute satisfactorily for heroin; methadone appears to block the heroin "high" and relieve the addict of his craving for it.

Dole and his colleagues are arriving at a comprehension of heroin addiction as a genuine epidemic disease. It certainly is highly communicable. It is spread by well-identified vectors. Among those vectors, please take note, there is not only the well-known "old dope peddler" but the entire apparatus of law and punishment society has created to suppress this vice. Our legislators, prosecutors, and police have no other counsel, from year to year, but to make the punishment fit the public's loathing of the crime. Increasing the severity of the punishment, however, increases the price of illicit drugs and thereby increases the profit of the trade and so ultimately promotes the spread of the disease.

The prognosis for this disease is poor. Heroin addiction, it is now realized, is the largest cause of death in this city in the age group afflicted, that is: the age group from fifteen to thirty-five. The reason you do not see many addicts older than age thirty-five is that they are almost all dead.

Defining the disease in this fashion—defining any disease in this fashion—surely extends the jurisdiction and responsibility of the physician far beyond the laboratory, the clinic, and the contents of his black bag.

The second affliction of the metropolis on our list is air pollution. Actually, no clear connection has been established between air pollution and health. Abel Wolman, the dean of the country's sanitation engineers, thinks it is principally an aesthetic affliction. Wolman observes: "If exhaust gases emitted by a Diesel bus had a fragrant aroma or, worse yet, led to physiological addiction not many people would complain about traffic fumes." Another close student of this subject, Eric Cassel, says we have here a cause but no disease to go with it. In Europe and the United States, the suspected role of air pollution in chronic respiratory disease is obscured by what Walsh McDermott calls "portable air pollution," that is, cigarette smoke.

Making its first approach to this situation, the Health Research Council has sponsored programs to monitor air at selected sites in the city. There is not yet nearly enough capital invested in the required instumentation at any point to observe all of the variables that are known to be involved or may be involved in the effects of air pollution on the human organism. The Health Research Council has sponsored also a promising investigation, under way at Cornell Medical College, that is using some highly sophisticated statistical techniques to uncover the correlation, if any, between the cycle of air pollution in this city and the rise and fall in morbidity and mortality rates.

The third affliction of the metropolis identified by the Health Research Council probably subsumes the first two and possibly all others that you could think of; this is medical care and its delivery.

The term "affliction" in this context refers, of course, to the inequities and inadequacies of the medical care system of the city. Three general statements can be made to mark the boundary of this empty territory on the map. In the first place, the magic-bullet age of medicine is behind us. Chronic disease and emotional disorder are the principal causes of morbidity in our population, especially among the 25 percent of the people who are on the downslope of life. Interns and residents in county hospitals in any big city of this country know that medically irremediable damage has long since been done to most of the patients above a certain age before they present themselves with some specific acute complaint. Paul Densen, the city's own house epidemiologist, has observed that neither the morbidity nor the mortality rates of chronic disease respond to medical care. It is apparent that prevention offers the primary route to the control of these rapidly enlarging concerns of the health of our population. We are compelled to say, therefore, that the task of medicine is no longer to cure disease but to secure the realization and fulfillment of the health and full physical capacity of all members of our society.

The second boundary for our *terra incognita* is established by the admission that the medical care available is not equitably available to everyone. New York City in the past generation has received a staggering influx of immigration from within the borders of the United States and from Puerto Rico. Yet, in this same period, the total population of this city has actually declined somewhat. The people, mainly white, who can afford to have fled the city to be replaced by poor people, mainly black, who think they are lucky to be here—at first, anyway, it looks better than the delta country in Mississippi.

The epidemiology of metropolitan medicine must learn to monitor the social and economic variables in the health of our people. Just eight blocks north of this Academy of Medicine building is 110th Street; north of 110th Street the infant mortality rate is double that recorded on the south. In Harlem and East Harlem, the life expectancy of the young adult is five to ten years less than that of the American population as a whole. Some 13,000 deaths a year are attributed to poverty and failure to deliver to the poor the care that modern medicine is capable of rendering.

The little that we know about the medical care system in New York City tells us, finally, that money cannot repair it. The system itself must be overhauled.

In the nation as a whole, over the past twenty-five years, the portion of gross national product devoted to medical care has gone up from less than 3.5 percent to more than 5 percent. Public expenditures for medical care have increased more than ten times and have increased, as a percentage of the total outlay for personal medical care, to more than 25 percent. The Medicare legislation now on the books, which is about to begin pouring more federal money into these channels, is surely going to increase the public percentage still further. Yet our country by no means leads the civilized nations of the world in the accepted indexes of health.

At this point, we should place at the center of our analysis the useful generalization made by Nora Piore, the foremost student of the New York medical economy. She enjoins planners to remember that the medical economy is always local. The national figures describe a Platonic abstraction that exists in no single community. Local variations above or below each of the averages and norms make them meaningless if they are to be applied as a guide to local action.

In the first complete accounting of the source and application of funds in the New York medical economy, conducted on her Career Scientist Award, Mrs. Piore has established that one-third of the bill for personal health services in this city is financed by public funds, federal, state, and local. The public $500 million, out of the total $1.5 billion for medical care, does not include the considerable public expenditures for medical education, research, or public health measures, nor the rising contribution from the public treasury to capital facilities.

This one-third of the funding of medical care pays the bills for 40 percent of the families of this city. Most of those families are otherwise solvent, able to take care of themselves and pay their bills. Thanks to the wisdom and humanity of the welfare system of our city, however, they are able to qualify as medical indigents. Elsewhere in the country, people have to certify themselves as plain indigents—paupers—in order to qualify for the flow of public funds in the support of medical care. The public contribution largely accounts for the fact that the expenditure per person for medical care in New York City runs considerably higher than in the country as a whole—$227 as against $142.

Here the celebration of the city's wisdom and humanity must stop. For the medical services that are financed by public funds in New York reach their clients through 25 different agencies; in a bewildering tangle of terms and conditions; with quintuplicate and septuplicate forms to be sworn to and notarized, and with discontinuity and fragmentation in the ultimate delivery of the care. In consequence, the care that reaches to the 40 percent of the families who depend upon public funds is care of the most expensive kind. It is medicine for repair of damage done; damage that is so often irreparable; damage that could have been prevented by expenditure of funds earlier and under more consistent economical, rational, and humane administration. It is inequitable enough that one-third of the total outlay should pay for 40 percent of the care. When it is realized that this is the most costly care, then what is delivered must be too little as well as too late. As Mrs. Piore says, the 40 percent get their medical care ". . . when they are very sick and very broke."

The Piore study closes the tiresome debate about whether medical care ought to be supported and administered through the public sector. There can no longer be any debate about public medicine in New York City, when public funds pay one-third of the total bill on behalf of 40 percent of the population. The debate must now turn to the more fruitful and challenging question of how to fashion a rational and a humane system of medical care out of the irrational and inhumane lack of system. Everyone will agree that we ought to use the massive flow of public funds effectively and efficiently, not only for the sake of the city and its medical dependents, but for the sake of the entire system. Park Avenue and Harlem have the same interests when public funds present the voluntary hospitals with more than one-third of their clientele and an increasing percentage of their income.

The bankruptcy of the local private fee-for-service medical economy is visible in the bankruptcy of the institution that was supposed to make it work: private health insurance. Some thirty years ago, S. S. Goldwater, then commissioner of hospitals, declared that the municipal hospital system was under intolerable pressure from the needs of the medically indigent—by which he meant otherwise solvent consumers priced out of the medical economy. He looked forward, he said, to the day when medical insurance would relieve the public institutions of this pressure. Well, the pressure remains intolerable. And it becomes the more intolerable as we turn up increasing evidence that the mounting public expenditures are not buying medical care of the right kind at the right time.

The Health Research Council hopes that this jungle, haunted with dangerous issues, will attract the concern and intelligence of the career scientists. There is a precedent for such venturing into the public forum by previously cloistered scientists. The physicists, twenty years ago, were alone in the discovery that nuclear weapons comprehend both the ends and the means of national power. They turned at once to the framing of recommendations for the control of their dread invention. Whether the ideas they advanced were politically feasible we shall never know, but for certain the scientists were right when they predicted that the old kind of politics, the old habits, and the old institutions would lead us into the dead-end street in which we find ourselves today. The design of public policy to secure the wise use of science must begin with those who are engaged in the advance of science. I would add: the innocence and directness of the moral attack that comes from the laboratory strikes more directly at the jugular of the new issues than all the sophistication from other quarters about what is possible, politically feasible, and acceptable to the consensus.

The success of modern medicine has plainly changed the terms upon which medicine can be practiced. On the one hand, the solo practitioner has outlived his time. E. D. Harrington of Jefferson Medical College has said: "More and more professional talent these days is being devoted to less and less serious illness among a smaller and smaller part of the population. The individual physician is selling convenience, availability and superficial personalization." At the other end of the spectrum, the general hospital, organized to handle short-stay acute afflictions by the depersonalized practice of high medical technology, is unequal to the task of coping with chronic disease and emotional disorder. Harrington says: "The greatest single danger in this combination of professional syndicalism on the one hand and open-market activity on the other is that it appears to be destroying a care system without offering anything like an adequate substitute."

Replicable models of the adequate substitute do not yet exist. The new institutions for the delivery of medical care will necessarily derive from those that exist today. Over the next few years, they are going to be fashioned by the squaring off, as always, of vested interest and special interest by compromise and negotiation and by all the modes of open and closed struggle known in American politics. The one interest not likely to find adequate representation is the public interest. It is the hope of the Health Research Council that the public interest in the creation of a rational system of medical care will find wise and determined partisans in the company of the Career Scientists.

...Getting Our
Knowledge
Fully Applied

FOR Lyndon B. Johnson the high point of his presidency may well have been the launching on June 15, 1966, of his Medicare and Medicaid programs. In an exuberant public appearance he envisioned the extension of the blessings of health and medicine to all citizens of the Great Society and went on to declare his own determination to see that they got the full benefits of the latest research: "Now, Presidents, in my judgment, need to show more interest in what the specific results of medical research are during their lifetime and during their administration. I am going to show an interest in results."

He disclosed that he had not reviewed his remarks in advance with his Secretary of Health, Education and Welfare or the Surgeon General and announced he was summoning them for what was to be the first of regular periodic conferences to review their progress in an "accelerated program of applying biomedical knowledge."

With the shock waves of this occasion still reverberating in the academies, Senator Fred Harris brought public officials, scientists, and other interested parties together in October in Oklahoma City for a stocktaking symposium to which I contributed this essay.

T_HE federal government in 1945, in the closing months of World War II, took over from the great private foundations the financing of science in America. A state paper published that year under the title "Science, the Endless Frontier" set out the premises for public policy in this new realm. In businesslike terms, it described science as the capital stock of industrial technology. Public outlays for science could be accounted as investment. The spectacular developments of the war years—including penicillin as well as radar and the atomic bomb—proved beyond all argument that science is a kind of investment that pays off. It followed that the Congress ought to build up the rate of public investment in science to a scale commensurate with the demonstrated promise of science.

The soaring curve of federal appropriations to what is now called "R and D" shows that the Congress has done its part. With the apparent approval of the electorate the annual outlay has increased from $500 million to more than $16 billion. There is some question as to just how much goes to science; that is: to the advance as distinguished from the application of knowledge. If the federal flow to the universities can be taken as an index, the figure exceeds $1.7 billion. Included in this figure is the more than $600 million laid out by the National Institutes of Health to the universities and their medical schools. The growth of federal support to the biomedical sciences has not lagged the total.

Since public support has been solicited on the promise of results, it is only reasonable to expect the public and its elected representatives to demand an occasional reckoning of results. This symposium has been called to consider a whole checklist of questions about results in the biomedical sciences posed by the President himself. The questions suggest no faltering of faith in science. On the contrary, the President declares his satisfaction that "a great deal of basic research has been done." In consequence, he asserts that "the time has now come to zero in on targets by getting our knowledge fully applied . . . to see what specific efforts can be made to reduce deaths among the leading killers, especially arteriosclerosis of the heart and brain and various forms of cancer, and to reduce disabilities such as arthritis and neurological diseases." He puts his most thorny question before us in the statement that, as compared to the "hundreds of millions of dollars spent on laboratory research . . . a very small percentage of research money is spent on research to [develop] new drugs and treatments on human beings."

Econometrics has not yet shown how to measure the worth of a dollar spent in the laboratory and compare it to another spent in the clinic. The public holds a stake in the advance as well as the application of the biomedical sciences. Since human and material resources are finite, even in our abundant economy, choices must be made.

The choices in this acutely difficult realm are, first and foremost, ethical choices. To individual human life we assign absolute value. We cannot fail to celebrate the triumph of compassion, nerve, and knowledge represented by the open-heart surgery that repairs the damage done by rheumatic fever and adds years to a patient's lifetime. But we must learn to find the same sense of triumph in the knowlege that such surgery will not have to be performed on men and women now in their childhood who are being protected by antibiotic drugs from the cardiac damage caused by rheumatic fever.

The value judgments that must be made in deploying public funds in the biomedical sciences call for a new kind of moral sensibility. C. H. Waddington has observed: "The most powerful forces operating in the world with which we have to grapple, intellectually and morally, are directed towards such ends as changes in the statistical indices of infant mortality or nutrition. The idea of the good, as we see it forcing the pace of historic change around us . . . can hardly be expressed except in terms of statistical parameters."

Walsh McDermott has summarized Waddington's observation. He calls for: "A statistical compassion . . . a true compassion for those we never get to see."

Medicine is now as big as agriculture in the American economy. Over the past twenty-five years the annual budget for medical care has increased from $8 billion to nearly $40 billion, from less than 3.5 percent of the gross national product to a full 6 percent. Public expenditures for medical care have increased ten times in this period, to more than 25 percent of the total. Over the next decade, it is predicted, the total outlay will increase to 8 or 9 percent of the gross national product, and the public outlays will then approach 50 percent of the total. Wisdom in the allocation of public funds on this scale will require the widespread inculcation of the capacity to feel and act upon "statistical compassion."

Let us try, therefore, to place the public stake in "an accelerated program of applying biomedical knowledge" in the perspective of our vital statistics. It is well known that the U.S. life expectancy has increased since 1900 from forty-seven to seventy years. The crude death rate has been all but cut in half, from 17.2 to 9.6 per 1,000 population (see graph, opposite). The age-adjusted death rate shows an even greater decrease, from 18 to 7.6 per 1,000 (see graph, p. 160). This summary index of age-specific death rates more faithfully expresses the true state of affairs because it offsets the effects of changing age composition; for example, the increase in the number of old people dying in a population with a larger number of old people.

Such figures are cited, again and again, in testimony to the "miracle of modern medicine." But the curve that plots these figures over the years from 1900 does not lend support to popular mythology. The curve for the age-adjusted death rate shows three distinct slopes—three phases in the story of progress—that remind us to distinguish always between medicine and health.

Over the first thirty-five years of this century, the crude death rate declined from 17.2 to 11 per 1,000. In this substantial progress—80 percent of the gain to date—medicine as we commonly think of it played a negligible role. Throughout the period, there was little the physician could do for his patient. It was, rather, the public health officer who made medicine's contribution to this improvement in the nation's health. The

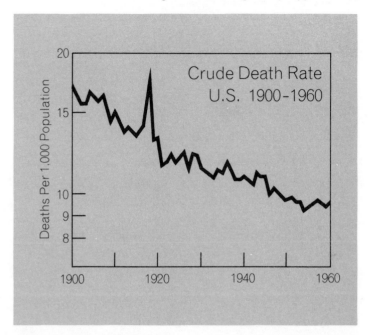

cleanup of water supplies, sewage disposal, food inspection, the installation of indoor plumbing, and other developments in modern urban technology called, in fact, for a still wider range of professional capacities. Ultimately, the vital statistics represent a summation of the nation's total economic growth and the increase in the general material well-being of the people.

The point is sharply illustrated by a statistical perception that we owe to Walsh McDermott. In 1900, 25 percent of the total death rate of the country was due to infant mortality. Of the infant deaths, half were caused by the so-called infant diarrheas and pneumonias. For these diseases, McDermott observes, medicine has no specific remedies, even today. Lord Boyd-Orr— founder of FAO and a physician who was led by the practice of medicine into politics—adds that medicine has equally little understanding of their causes, beyond the recognition that "the one is caused by bad feeding and the other by bad housing." They continue to cause half of the infant deaths that contribute so heavily to the high death rates of the poor countries of the

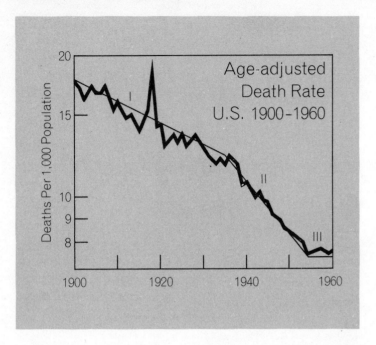

world. In our country, these diseases and deaths have dwindled away. They declined with the birth rate, in the first place. Their decline coincides also with the percolation of general well-being down through the income groups to the point where these diseases are isolated in the enclaves of the most wretched of our urban and rural slums: "Out of sight, out of mind."

To the second phase of improvement in the nation's health, modern medicine—administered through the physician-patient relationship—has made its indubitable contribution. Between 1935 and 1955 the crude death rate declined from 11 to 9.6 per 1,000 population. If the drop in the crude death rate does not seem to be a large one, the new power of medicine shows up plainly in the slope of the age-adjusted death rate. From the average fall of .15 percent per annum prior to 1935, the curve steepens by 60 percent to an average decline of .22 percent per annum from 1935 to 1955 (see graph above). During this period, the antibiotics eliminated the infectious diseases from first rank among the causes of death in the United States.

There were, of course, other victories, such as the polio vaccines. In the slope of the crude death rate, the elimination of polio shows up hardly at all. Yet, to the parents of children growing up in this period, the gain looks larger than all of the 80 percent decrease in the death rate accomplished before 1935.

In the third and present phase of the evolution of our vital statistics in this century, the death rate has flattened out. For the past decade it has run parallel to the base line, hovering within one or two tenths of a percent around the average of 9.6 per 1,000. The oscillations reflect the smoldering aftermath of the worldwide flu epidemic that caused the sharp peaking of the death rate in 1918 and tell us that the vaunted conquest of infectious disease is not yet complete.

The reduction of the infectious diseases has brought the organic afflictions of the human body to the fore. The American people live long enough to die of old age. The diseases of the heart and circulatory system and cancer—broadly speaking the degenerative diseases associated with the aging of the organism—account for more than half of all the deaths in the U.S. population. They stubbornly resist the clinician and the investigator alike.

The little progress that has been made in the treatment of these diseases, however, sustains the public expectation of science as well as the hope of individuals. "Heart disease, cancer, and stroke" have been delimited as a new frontier of American life. Federal legislation under this title in 1965 summoned the medical profession to place emphasis on education, clinical research, the training of clinicians, and on experiments and demonstrations of clinical care of patients on a regional basis. Far-reaching effects upon the institutional structure of American medicine are anticipated as this legislation takes effect, especially upon the relations between the medical schools and the profession at large, between the gown and the town of medicine. The cost of the program has not yet been calculated, however, and the benefit—whether it will have appreciable effect upon the death rate—cannot be predicted.

A person, and even a people, can be forgiven for wanting to live forever. Nonetheless, one must consider whether, at 9.6 per

1,000 population, the U.S. death rate is not approaching an irreducible limit. The answer is that there is room for progress, perhaps for as much progress as was won beween 1935 and 1955.

Despite the huge absolute and relative expenditure for medical care, the United States falls behind a number of industrial countries in aggregate death rates and behind many national communities in age-specific rates. A comparison of the U.S. experience with the lowest recorded age-specific death rates abroad, undertaken by the National Center for Health Statistics, suggests that this country might be able to reduce its crude death rate to as low as 7.3 per 1,000 population.

More detailed comparisons of this kind, from point to point on the age-specific curves, yield insight into the reasons why the United States trails other nations in attainment of objectives that it prizes so highly. Only in the eighty-five and older age group does the U.S. record compare favorably with others. It comes as a larger shock to discover that the biggest gap between the United States and other civilized countries appears in the infant death rates. If one then consults the interior detail of the U.S. vital statistics the factors underlying this discrepancy stand forth plainly. The infant death rates for the nonwhite population, wherever they are recorded across the country, run twice that of the white.

Since color is so inextricably identified with poverty, the discrepancy may safely be taken to reflect socioeconomic causes. That surmise is supported in those few instances in which health records are correlated with income and occupation. A study for the Health Insurance Plan of Greater New York shows, for example, direct correlation between high infant death rate and low quality of medical care. The infant death rate for the white population, in other words, is elevated by the unfavorable experience of the white poor.

If the higher mortality rates from all causes that prevail among the nonwhites may be taken as reflecting the neglects and disadvantages associated with poverty, then the strategy for reducing the number of deaths in the United States population—if not for reducing the toll of the leading killers—becomes plain. Subtraction of the poverty-specific death rates from the U.S. totals would bring the U.S. crude death rate down close to that figure of 7.3 per 1,000 population, synthesized from the best experience elsewhere in the world.

Happily, the administration and the Congress have already committed the nation to abolition of the economic barriers that stand between so many Americans and the medical care they need. The 1965 amendments to the Social Security Act are designed to make human need effective as economic demand. People who would otherwise receive medical care as objects of charity or not at all are now to be enabled to present themselves as fully certified consumers. The first 19 million, those entitled to social security annuities, were automatically qualified by Part A of Medicare. It is an extraordinary fact that by the first of July more than 90 percent of these people, nearly 18 million, had qualified themselves for the "voluntary" benefits extended by Part B.

If Medicare stood alone, it would not reduce the economic barriers, even for the 10 percent of the population that it serves. As an insurance measure, it interposes a whole set of limitations and deductibles between the beneficiary and the care he needs. It is on the poor, who stand in greatest need of this protection, that the limitations and deductibles fall with the effect of cruelly regressive taxation. The inequities here are redressed in part by Title XIX. As a welfare measure, this legislation extends federal aid to pay for care rendered to anyone, no matter what his age, who cannot pay his medical bills.

Title XIX, which New Yorkers have labeled Medicaid, may prove far more significant than Medicare. Under its terms, the condition of medical indigency is recognized for the first time in national legislation. It is left to the states to define that disability as it locally prevails.

The prospective impact of Medicaid upon the nation as a whole can be forecast best from the experience of New York City. For reasons special to the city's history as the country's melting pot, as an early center of great wealth and philanthropy, and, perhaps above all, as the capital seat of medicine in America, New York long ago incorporated the notion of medical indigency into its municipal constitution. Well before the Congress offered the present subsidies, public funds were paying a significant portion of the cost of personal medical care in New York City. The clientele includes the city's welfare families, but it embraces a much larger number of families headed by persons gainfully employed and otherwise able to pay their bills. The medical care they receive—fragmented, discontinuous, depersonalized—is the most expensive kind of care; 70 percent of the outlay is for hospital services.

The New York model—with annual expenditures *per caput* increased by public funds a full 60 percent above the national average—shows that money alone cannot produce good medical care. It is becoming clear that there is a serious mismatch between the system for delivery of medical care and the real needs of the American people, now that those needs can be stated in dollars.

The failure to make connections is implicit in the expressions of disquiet that are heard in presumably responsible quarters about the future training of surgeons. If the new arrangements are to eliminate wards and ward patients, it is asked, where is the necessary teaching material to be found?

Of course, the teaching subject often receives better care than anyone can buy outside a teaching hospital. Yet the question lays bare the moral void at the heart of the dual system, the double standard, of American medicine. Ability to pay for medical care bears no natural connection to the care needed by the person in question. All the vital statistics insist that the relationship is invariably inverse. Here is the point of ethical disjunction. It is precisely here that the system is disjoined from the needs it is bound to serve in the future.

The system of medical care prevailing in America embodies, above all, the satisfaction of its beneficiaries on the two sides of the fee-for-service, free-choice-of-physician arrangement. The system is wonderfully designed to manage acute, episodic, and unusual diseases and injuries. Understandably, crises of disease and injury take absolute precedence in the consumer's budget; for the worst occasions the best-off are ready to spend the most. Correspondingly, among the members of the highest paid of all the learned professions, the most richly rewarded are the physicians and surgeons trained to cope with the severest and rarest of afflictions.

In order to prepare the medical student for the opportunity of such service, the medical schools in turn concentrate on acute, episodic, and unusual diseases and injuries. The research budget and effort inevitably tend to follow the same pattern. As Willard C. Rappleye has observed, the afflictions that are the object of this outlay of treasure and talent "constitute only a small segment of the total spectrum of health problems."

The sensitive senior student also takes note that the interesting cases on the ward are vastly outnumbered by the "crocks." In preparation for the rough and tumble of competition in the market, however, he can tackle only a small segment of the total spectrum.

Without doubt, the system that functions under these extremes of social and ecomomic pressure has produced care of extraordinary perfection and efficacy for those portions of the population that it serves and reaches. When it comes to vital statistics, however, the evidence indicates that this system has not played as significant a role in the general improvement of the public's health as the overall steady growth of the economy as a whole.

Now that Medicare and Medicaid are to lend economic sanction to human need, the medical care delivery system must give priority to new and neglected services. With federal funds to share the burden already carried by the state and city treasuries, New York City may pioneer replicable models for other communities in the nation. The city's new Health Services Administration is seeking the functional coordination of the institutions in the private and public sectors to provide community-based health care tied to the teaching hospitals and the six medical schools in the city. Against the high infant and maternal death rates in the slums the new system will have to pioneer the development of educational as well as preventive services. For people at the other end of life, it must provide facilities and services that do not now exist for the care of chronic disease. Adequate response to the demand generated by

Medicare and Medicaid for preventive and ambulatory services could empty one-third of the beds in the general hospitals whose occupants do not belong there.

If medical care of the quality now available to the most fortunate 40 percent of our population can be made accessible to the people at large, the deaths and disabilities caused by the "leading killers" and their runners-up can be substantially reduced within a decade. Recalling the important distinction between health and medicine, we can count upon a substantial assist toward this objective by the 10 percent increase in the social security benefits proposed by the President to begin next year. In sum, the 1965 amendments to the Social Security Act promise the most effective way to accelerate the application of biomedical knowledge.

Without doubt there are also significant gains to be made by accelerating the movement of new knowledge from the laboratory to the clinic. John M. Russell quotes a medical school dean as saying, "Now if a child with cystic fibrosis happens to live in Boston, he gets superb services there. If he lives in another part of the country near a medical center, his services might be routine because the pediatrics department there might be principally interested in the newborn infant or in mental retardation." With the encouragement of the health legislation enacted by Congress in 1965 and with the support of federal funds, Russell concludes, "this sort of thing will not happen much longer."

There are hazards, of course, in the process of acceleration. One is posed by the already high speed with which new drugs and treatments get into circulation. The schools and the professions as well as the government agencies are on the alert, however, and the United States did escape the thalidomide catastrophe.

There is another more pervasive and intangible hazard in the climate of opinion enveloping the country's medical establishment. No one has suggested that the research effort be reduced in favor of application, and the flow of research funds from Bethesda has not diminished. But the twenty-year upward trend in research appropriations has halted, and to a certain extent the new programs threaten to absorb new funds that would otherwise go to research. It can be argued, no doubt, that an occasional fiscal squeeze of this kind is a good thing for all concerned. The project contract/grant system has tended to set up positive feedback loops that have amplified some lines of work out of reasonable relationship to the whole. Throughout this period, however, the federal investment in biomedical research has borne about the same percentage relationship to the national medical economy as the total R and D budget to the gross national product. If the medical system is to grow soundly toward the predicted 8 or 9 percent of the gross national product, the research outlays must grow with them.

Beyond what may be accomplished by the introduction of equity into the delivery of medical care, large increases in longevity and significant decrease in the mortality rate can come in the future only as the result of a general breakthrough on the broad front of human development and aging. The cardiovascular diseases and cancer involve the deepest questions man can ask about the nature of his existence. Correspondingly, the biomedical sciences seek answers that go far beyond "results." Research in these realms claims public support because it enlarges life's meaning and the purposes for which men live.

An End to
Welfare Medicine

IN 1966, as President Johnson was celebrating
the establishment of Medicare, John V. Lindsay was confronted
with what appeared to be the ultimate breakdown of the municipal hospital system of New York City. At the mayor's behest
I undertook to organize a commission of private citizens to
consider this problem in the context of the still larger problem
presented by the unsatisfactory performance of the entire system for delivery of personal health services in the city. The
following essay, prepared for the commencement of the Class
of 1968 of the Cornell University Medical College, presents the
gist of our report.

I F THERE is such a thing as a doctor in the generalized pluripotent, omnicompetent medical sense of that title,
it is you in your caps and gowns here today. Four and a half
years from now you will each have mastered some branch,
corner, or aspect of the technology of medicine, and you will

have differentiated into one or another of an ever increasing variety of specialists. From this it follows that the hallowed image of the one-to-one, face-to-face physician-patient relationship no longer fits the reality of medical care.

The availability to any patient of the full resources of modern medicine requires the mobilization and coordination of the training and experience of many more than one physician. We would have to reassemble your entire class, with perhaps two or three specialisms in surplus, and would have to reach into other classes and perhaps other schools to fill in the gaps. In addition we would have to enlist still other technical and supporting personnel.

The practice of medicine now even involves people like me, who have been working at the organization of the social and economic institutions and arrangements needed to put twentieth-century medical technology to work. It is on this intractable topic that you want to hear from me.

I shall tell you, therefore, what I am going to tell you: The market process cannot effectively organize the technology of medicine for service to the individual and the community. The creation of the necessary new institutions and arrangements must begin with the complete reconstruction of the relations between the public and the private sectors in medicine, between public authority and voluntary initiative. The hope for a rational and optimal outcome hangs heavily upon the intellectual leadership of the medical profession. In brief, I am here to talk to you about another task for the medical schools.

Needless to say, as we proceed to consider what might be desirable, events are already in motion, outside of any deliberate or rational control. The fat is in the political fire. This is because the Social Security Amendments of 1965—the Medicare and Medicaid legislation—enacted into law the worst features of the prevailing arrangements.

That legislation dumped hundreds of millions of dollars into the fee-for-service, solo-practice market economy. The insurance feature of the legislation, in the established pattern of medical insurance, places obstacles in the way of preventive—that is, accessible and early—care and promotes resort to the most costly modes of care, to hospitalization and surgery. The welfare feature exerts the same economic bias in favor of the acute episodic care and interposes an obnoxious means test between the patient and the care he needs.

Within a year, the surfacing of unmet needs and the inflation of prices in the medical marketplace had brought such a huge excess of expense over estimates that the folk economists in Washington and Albany were moved to cut back on the benefits provided in the original legislation and to reduce the availability of what remained. The reverberation from that welching on a long-overdue public undertaking is now moving from the legislature into public consciousness. Former Medicaid beneficiaries are once again being rejected by the voluntary hospitals and sent back "where they came from"—to the municipal hospitals. We may assume that these beneficiaries are formulating their own position as to human wants and fiscal priorities. This is sure to be an issue in the next cold winter if not in the long hot summer just beginning.

We scarcely needed the present crisis to demonstrate the failure of the market. Even in the bad old days, when medicine could not do its patients much good, the theoretically self-regulating economic mechanism did not secure the delivery of care to all in the community. Charity filled the gaps. The modern hospital had its origin as a refuge for the poor in the days when the rich died at home. The open ward was the standard accommodation, and the private room did not figure in the design until medicine and surgery had services to offer that required the workshop we now know as a hospital.

Under the dual system of medical care, conscientious physicians doubtless gave as much devotion to their charity practice as to their income-yielding patients. Caste distinction was nonetheless imposed by disparity in the physical circumstances in which care was rendered and by the moral circumstance which submitted the charity patient to treatment as teaching and research "material."

With the cost of medical care increasing along with its effectiveness, charity has yielded to welfare in the financing of the delivery of care to the medically indigent. Public funds now, therefore, go to perpetuate the caste system in medical care. Under the terms of the present arrangement between the government and the private economy of medicine, public authority is limited to filling the gaps.

Here in New York City we behold the caste system in medical care working at its optimum. The patriciate of this city—meaning the grandsons and -daughters of the immigration wave before last—has always had a close working understanding with the leadership of the latest wave of immigrants. In consequence, the city of New York has historically devoted a larger percentage of its income to social services than has any other big city in our country. Instead of the single giant county hospital that represents the public sector in most American communities, the city of New York operates a chain of twenty-one municipal hospitals—one-third of all the beds in town—plus a patchwork of clinics and neighborhood services. For many years before Washington came into the picture, the city paid for half of the patient days in all the hospitals; for the delivery of nearly half of the babies; for such pediatric services as were available to half of all the children; for the in-hospital and nursing-home care of perhaps 80 percent of the elderly. Nor was there stinting of expenditure; the city's outlay for its medical dependents exceeded the national average expenditure per capita.

The most recent appraisal of the working of this system is that produced by the Commission on the Delivery of Personal Health Services which Mayor Lindsay asked me to organize in December, 1966. Our findings, after a year of study, may be summed up in the declaration that the services provided are substandard but that the services not provided—the gaps that remain unfilled—constitute the real hole in the health services, public and private, in this community.

First-rate professional services are available, under the most wretched moral and physical conditions, in the city's own hospitals and, under better conditions, in the larger voluntary hospitals that render much of the care paid for by the city. For lack of ambulatory-care facilities and services, however, the medical problems presented at the city hospitals and in the wards of the voluntary hospitals are far more acute and difficult than those encountered in private and semiprivate accommodations. (That is why the poor make such good teaching material.) For lack of adequate chronic-care facilities, the elderly crowd the wards of the city hospitals and the emotionally ill are shipped upstate to custodial care in state hospitals.

It should be understood that the failures charged here must be laid not to the city alone but to the private medical economy as well. Allocation of resources in accord with the distribution of effective demand leaves gaps. The city is unable to outbid the competition for the resources it needs to fill those gaps it tries to fill.

The failure of the health services of New York City affects the welfare not only of the poor—not just the city's dependents on relief—but of nearly half the population. There was much misunderstanding and misdirected outrage at the $6,000 annual income standard that qualified a family of four for services under the original Medicaid legislation passed in Albany. In point of fact, the health services of the city of New York had long recognized a somewhat higher threshold of medical indigency. The threshold tended to be set by pragmatic humanity, not by welfare accountants.

The sort of person who insists on stricter criteria might find the facts uncomfortable. Americans devote more than 6 percent of the gross national product to the provision of personal health services. The national average expenditure now stands at $200 per capita. A little arithmetic shows that for a family of four such expenditure as a fair charge to income requires a family income of about $15,000 before taxes. That income is enjoyed by less than 15 percent of this country's families. This is not to say that 85 percent of the population are medically indigent. Not that many are seeking the care, especially the preventive or "maintenance care," they ought to be getting at all times. But it is a fair statement that 85 percent are at all times exposed to the hazard of medical indigency and can be reduced to general indigency by a sufficiently catastrophic experience.

For the population of New York City, the great Indian gift of Medicaid has, without doubt, raised the threshold of medical indigency higher than before. A visitor to a city hospital emergency room is now confronted with a bill for $11. It may be that no one insists the bill be paid. But the bill itself sets up a serious psychological hurdle against a return visit. The infant had better be really sick the next time to justify such anxiety and humiliation for his mother. No such transaction compromised municipal medicine in New York City, for all its deficiencies, until Washington and Albany endowed us with Medicaid.

The conclusion that the market cannot organize a technology as complex as medicine should come as no surprise to anyone who has a little familiarity with the U.S. industrial system. That system does not leave the organization and operation of any other high technology to the play of market prices. The steel industry does not haggle with an iron-ore industry. It has integrated its technology from mine to warehouse and has carried its integration forward to deliver about one-quarter of its total product in finished form to final markets. The same genius for logistics makes it possible for the auto industry to turn out 10 million cars a year.

It is not left to the consumer to assemble his own automobile from this manufacturer's motors and that manufacturer's fenders. By the same token, the consumer—medically indigent or otherwise—can no longer be expected to organize the medical care he needs out of the highly specialized working parts available to him here and there in the present system.

It was reflections on this line that led the Commission on the Delivery of Personal Health Services to its primary recommendation. This is the recommendation that the city should get out of the hospital business—that the city should stop trying to fill gaps in the private delivery system with its own inadequate facilities. Instead, we recommended, the city should undertake to secure the delivery of personal health services to the people by contract. The city should use the legal instrument of the contract to enlist the private sector—the voluntary institutions that command the resources of modern medicine—in the task of making health services available to all on the basis of need and not on any test of ability to pay.

In fact, the city has already contracted out about half of its total burden in the delivery of health services. Before Medicaid, the city was spending half its health-service funds to enlist the resources of voluntary hospitals and medical schools. Through the charitable institutions' budget of the Welfare Department, it paid for a considerable volume of services rendered by the voluntary hospitals. Through so-called affiliation contracts, the city hired voluntary hospitals and medical schools to staff the services in its own hospitals. With Medicaid and Medicare funds available to extend city tax funds, a still larger portion of the burden of tax-subsidized health services has been transferred to voluntary institutions.

In effect, therefore, we were recommending that the city contract out 100 percent of its commitment to provide medical care for all who need it. To get itself out of the health service business we suggested that the city turn its facilities over to a nonprofit Health Services Corporation. The corporation would have two functions. It would, first of all, disengage the twenty-one city hospitals from the suffocating constraints of the "checks and balances" system of the city government. That system, set up by generations of well-intentioned reform administrations, is supposed to keep the municipal government honest. Though there is no proof that it prevents petty larceny and every indication that it is vulnerable to grand larceny in broad daylight, the system most effectively promotes waste and delay, frustration and procrastination, incompetence and impotence at every level and in every department. Disengagement of the city hospitals from city management would be a first step toward erasure of the caste distinctions that now so invidiously segregate them from the voluntary hospitals. The Health Services Corporation would have a second equally important function. That is to deploy the city's considerable assets to promote the rational structuring of the entire health service system of the community—to give the system a structure appropriate to modern medical technology and responsive to human needs.

Flow charts and organization tables for the integration of the highly differentiated and specialized working parts of the technological apparatus of medicine have been on the drawing boards since the first Hill-Burton hearings at the end of the Second World War. It is here that the medical schools are called upon to assume a major new function in our society. These idealized systems invariably place a medical school and its teaching hospital at their peak or center. With six medical schools and a dozen first-rank teaching hospitals in New York City, there are more than enough resources to give each borough its own regional health-services system. The design should turn on people, however, not institutions or political jurisdictions. A fully comprehensive system of health services places the individual at its center; from the moment he enters a doctor's office or other site of primary care, he is within reach of whatever resources he may need.

The operation of the laissez-faire economy in medicine has given us a poverty of information about the relationship of resources to needs. There is now, for example, no reliable basis for saying how many physicians with this or that competence are required to serve a given unit of urban population. This is a question that ought to be of considerable interest to medical schools, and one of many about which they are bound to learn more in their new relationship to the community.

Accepting prevailing availabilities, it appears that 1,000 people will keep a physician employed; 30,000 to 50,000 people will fully occupy the talents of 30 to 50 physicians representing the major specialties in a well-staffed ambulatory care or group-practice clinic; 200,000 to 300,000 people will keep the attending and resident staff of a 300- to 400-bed general hospital occupied; while a population of 1.5 to 2 million people requires the full range of services to be found in a major teaching hospital or medical school. Each ambulatory care center, group practice unit, or individual physician in such a system has ties to a community hospital; the community hospitals are tied in turn to the medical center of the region. Informal channels of communication and referral already link up institutions in this city and, indeed, tie the medical centers of the city to hospitals and physicians throughout the New York metropolitan region, with its population of 15 to 20 million people. For the purpose of securing responsiveness of resources to needs, these communities of affiliation must be made explicit and should be formalized by contract.

One important objective is to resolve the clash of mission—

between service on the one hand and teaching and research on the other—now besetting the medical schools in this city that have taken on responsibility for the services in municipal hospitals under the affiliation contracts. With the municipal hospital system once again the sink for patients rejected by the voluntary system, those medical schools are confronted with open-ended liabilities. In properly structured regional federations, the liabilities would be equitably shared by the participating institutions. Each community hospital would be obliged to receive and care for anyone in its community requiring its service. The medical school, in charge of the regional medical center and backstopping the community hospitals with its command of the full spectrum of medical technology, would carry a selective burden of service appropriate to its teaching and research missions.

A fully comprehensive system lays as much emphasis upon preventive and supportive measures as upon treatment for illness and injury. In the vivid language of Martin Cherkasky, chief of Montefiore Hospital and head of the new department of community medicine at Albert Einstein College of Medicine, the object is "to narrow the front door of the hospital and widen the back door." The rising costs of medical care require this rationalization of the precious resources of medicine. Some portion of the current inflation in the medical economy can be charged to upgrading of wage scales that have hitherto made nonprofessional hospital employees serve as involuntary philanthropists. Most of the cost increases, however, reflect the unconscionable waste of resources encouraged by present modes of compensating and reimbursing physicians and institutions. In a structured, regional system it becomes possible to establish economic and social incentives for more efficient use of resources.

The role of public authority in the delivery of health services now becomes self-defining. As the contractor purchasing health services for the citizenry and, in the process, supplying the major portion of the capital requirements and operating income of the health-services system, the city government must qualify itself to write and secure fulfillment of the health service contract. The city must create an entirely new kind of health agency, charged with assignments not now covered by any public agency at any level in the government.

The first of these assignments is to monitor the health needs of the population, to practice a new kind of epidemiology that fixes the denominators defining the public needs. On the basis of this intelligence, the health agency would plan the allocation of resources. Planning has been specified as a governmental function in both federal and state legislation; it is the decisive power in the framing and implementation of public policy. Its wise exercise calls for the closest collaboration of public and voluntary agencies.

Through its contracts with the providers of services, the city health agency would secure fulfillment of the plan. Its contract with the Health Services Corporation, in particular, would give the agency leverage on the entire medical economy, to encourage the integration of its parts into a comprehensive community health service system. Against the inventory of public need and the declaration of public policy set out in the plan and contract, the agency would, finally, audit the performance of the providers, securing their accountability for the use of public funds.

In establishing the Health Services Administration in the government of New York City, John Lindsay and his colleagues have laid the foundation for the creation of this new kind of health agency. Responsibility for health services is now lodged for the first time in a single office, instead of being allocated haphazardly among a half-dozen departments and mostly lost in the cracks and crevices between them. With the Health Services Administration in being, it becomes possible to envision an entirely new kind of partnership between public authority and voluntary initiative. Through the instrument of the contract, each can make the contribution it is best qualified to make to the organization and delivery of health services.

A combination of despair and hope persuades me that the next steps will follow soon. On the one hand, neglect of public need has been compounded by betrayal of public expectation; the political pressure gauge is now reading in the red zone. On the other hand, it is apparent, all six medical schools in this city are ready to embrace the responsibilities that the progress of their technology has thrust upon them.

V Too Many
Indians

The Grain That
Steel Might Grow

THE late Abraham Stone, who with his wife Hannah directed the Margaret Sanger Research Bureau in New York City, responded to the welcoming committee on his first visit to India in 1949: "You asked us for bread, and we have sent you a stone." This gentle, self-deprecating joke now, two decades later, becomes a damning indictment of the abuse by the United States of its wealth and power in the world: they asked us for technical and economic assistance and we sent them helicopter gunships, high-explosive bombs, CS gas, napalm, and defoliants.

In the four essays in this section I have attempted to draw the lines of perspective which, if reason prevailed, would bring the foreign policy of the United States into more constructive connection to the natural history of the human species. In the perspective of biology, the most important current event in that history is the increase in the rate of increase of the world population. This increase comes principally from the population growth of the poor or developing countries. In the perspective of cultural evolution, this sudden multiplication of the poor signifies at once the beginning and the slow pace of the industrial revolution in those countries. The apparent contradiction in this statement is explained by reference to the history of the countries that have had their industrial revolutions. Death control, in each case, accelerated their population growth at the outset; later, as entire populations experienced increasing well-being, fertility control brought the rate of growth downward and, in some cases, even brought population growth to a halt. In the nearer perspective of this unprecedented revolution in demography, I conclude that the rich countries should place substantial resources behind the industrialization of the poor countries. The more rapid their industrialization and the more sudden the attendant population explosion, the sooner can the species set a smaller and more manageable ultimate limit on its numbers.

There is more to industrial revolution and population control than is expressed in this bare technocratic equation. In any case, the foreign policy of the United States has had a different theme during this period, quite irrelevant to natural history. Economic assistance to the developing countries, or "foreign aid," never acquired political leverage. It has even less electoral appeal now than at the beginning of the decade when the momentum of Marshall Plan aid to the country's cold-war allies was carrying economic assistance to the Third World. The connection between population control and the pill or the intra-uterine device is easier to trace than the more roundabout logic that places a rolling mill or a computer at the center of the picture. Even at its peak and at its best, the "AID" program was burdened with a condescension that gave second priority to capital inputs in favor of consumption commodities. The only political interest to rally behind the program was the farm lobby. It was the Soviet Union that built the steel plant at Bokaro, India, on which the United States reneged, as recited in the first essay in this part. The AID appropriations have continued in steady decline into the current fiscal year and are even more completely dominated by purely military-strategic objectives, with motivation by any larger and longer-range concerns vanishingly small.

Now that Americans are enjoying the luxury of misgiving about their own material abundance, it is probably the more impossible to interest them in assisting the industrial development of others. Meanwhile, the world-wide industrial revolution is proceeding without us and despite us. From Latin America, black Africa, and the Middle East as well as from Indochina come signs that the revolution may soon turn against us. The four essays in this section were written with more hope in 1964–1966; they have relevance, I believe, to consideration of the national interest today.

The first of these essays I delivered in response to the bestowal of the UNESCO Kalinga prize on me in New Delhi in January, 1964.

T HE map of the world shows two kinds of nations: "developed" and "underdeveloped," or, in plain language, rich and poor. The rich nations embrace less than one-third of the world population and consume more than two-thirds of the world's output. To most of their citizens the rising productivity of industrial technology furnishes increasing well-being; in some of these countries, surpluses have begun to embarrass the economic order. In the poorer nations, on the other hand, the output of preindustrial agricultural technology tends to fall constantly behind growth of population, and the material circumstances of most of their citizens are declining. Of course, poverty is not new to the experience of the mass of mankind; the rich nation is the innovation of recent history. Today, however, the poor are politically mobilized, and the new nations of the poor are committed to industrial revolution. For a century to come, the development of the underdeveloped countries will set the course of history.

No physical barrier stands in the way of this culmination of the industrial revolution. An exhaustive and authoritative accounting—rendered by scientists of many nations and by the international civil service, now two generations old, of the League of Nations and the United Nations Organization—

assures us that the earth's resources are ample to the needs of a much larger world population. Existing technology is equal to the task of accelerating the increase of production ahead of the growth of population everywhere in the world. With the consequent improvement in individual circumstances, there is good reason to expect that populations will stabilize well within the capacity of resources and technology to provide material well-being for all. The birth rate has followed the death rate downward in every country which has experienced development.

The only uncertain question is the process of development itself. Development is not an exercise in economic theory; it is an enterprise which engages the energies, ambitions, and passions of men. The open question is the cost of development—the only cost which counts: the cost to the liberty and life of the living generation of men.

On the precedents of the past, reaffirmed by each cycle of development, the prospective cost of the next round—multiplied by 2 billion lives—makes the heart contract. To say that the question is open is to declare a hope. The hope is that history need not go on repeating itself.

Because history is written so largely by its beneficiaries, history has little to say about its human cost. The archaeological record shows, however, the invention of slavery soon after the agricultural revolution and before the writing of history began. This primitive economic institution worked well enough to lift civilization high and carry history forward to great triumphs and tragedies, all on the basis of quite primitive technologies. Out of the always insufficient and relatively constant product—per man-hour and per acre—it secured a sufficiently inequitable distribution of goods to support a minority in occupations which called for the exercise of higher human capacity than the scraping of brute existence from the soil.

Over millennia, development was imperceptible. The same treadmills and capstans are to be seen in the friezes of Egyptian tombs and in engravings of the seventeenth-century London dockside and are still at work today in the fields and on the wharves of India. All at once, for reasons not known, the turn in history came from the West. Not long after Galileo and

Newton had secured the foundations of celestial mechanics, the rate of technological change quickened, and the mastery of mechanical forces began to amplify the strength and skill of men. The surplus so long extracted from scarcity by deprivation of the mass of the people found a new social function. It became the capital for increasing the productive capacity of society. The historians of the first industrial revolution are just now recognizing, however, that the process of capital formation involved a sharpening of the inequities that produced the surplus. The "savings" that financed the classical industrial revolution of England were principally involuntary; that is, the savings were taken from the reduction of the material estate of the yeomanry. This was the era of "carboniferous capitalism"; Asa Briggs has shrewdly traced the institution of the "two cultures" to the aesthetic and social protest of contemporary men of letters against such ugly manifestations of applied science as the slag heap and the mill prison. The coercion and cruelty, the pain and rebellion, of those days are largely forgotten or are healed in the amenities enjoyed by the latter-day heirs of the first cycle of development. E. P. Thompson closes his history of the English working class with the prayer that "Causes which were lost in England might, in Asia and Africa, yet be won."

The story has been the same in each industrial revolution that has followed. In the United States, historians continue to celebrate the frontier. But it was 35 million steerage immigrants—a flood of humanity equal to the nation's population at the end of the Civil War—who furnished the primary capital for the industrial revolution which got under way at the middle of the nineteenth century. The American Negro's present determined drive to capture and assert his civil rights, after a century of putative emancipation, serves to remind Americans how the savings for the development of their country were corraled.

Meanwhile, the nation which has most lately joined the circle of the rich nations looks back in dismay on the cruel human cost of its brief, ruthless, and successful revolution. Its moralists and economists deny that the now acknowledged costs are inherent in the dictatorship of the proletariat and argue rather that these were paid to foreign invasion, counterrevolution, and the cult of personality. Presumably, another generation of economic studies will find that coercion served here, as it has served elsewhere so invariably, the elementary function of capital formation.

If these precedents must be followed, then the next few years will see the developing nations, one after another, come under authoritarian leadership. Admittedly, ad hoc regimes are more common than not among the poor nations, regardless of their rate of development or degree of stagnation. The nations which are moving, however, are those which are most harshly governed, and the rate of development seems to correlate directly with the disposition to apply and submit to coercion.

Against the example of history past and present, India stands alone. India's leaders have declared their determination to bring about the industrialization of their country and the attendant radical reconstruction of the social order through the institutions of political democracy and without resort to coercion or the invocation of class hatred and violence. This experiment in development is as crucial as it is unprecedented. Should India fail, the grim cycle is fated to go on as before. Nor can the rich nations hope to remain spectators. Those in particular which hold stakes in the one-crop and mineral-extracting economies of the preindustrial poor nations must anticipate that the fervor of the new nationalisms will be turned against them. Given the instability and disorder in the relations of the rich nations, the proliferation of authoritarian regimes among the poor will constantly increase the danger of war. That danger must ultimately become intolerable, as thermonuclear weapons find their way into the armament of the "nth country."

If the Indian experiment succeeds, on the other hand, this could be as significant a turn in history as the industrial revolution itself.

It is not yet possible to forecast either success or failure. The percentage gains under the first two Five-Year Plans have been great. But the absolute magnitudes are small compared with the unmet need. One can say bravely that the work has begun; but it is also clear that time is running faster with each month and year. The rapid increase in India's urban population—the urbanization of poverty barely endurable in the shelter of the village—must go to destabilize political institutions not yet secure.

The figures for the first two Plans show that their execution fell short of their goals; the Third Plan lags its timetable as well. But the figures show another fact: the gap in the Plans correlates with a persistent shortfall in the aid projected from overseas. After three Plans, the lesson should at last be clear. India's vision of peaceful development rests on the expectation that the rich nations of the world will supply—by grant or long-term loan—a portion of the necessary capital. In the Third Plan this portion is explicitly declared to be 20 percent of the total investment program.

By what right or logic does one nation thus lay unilateral claim on the wealth and bounty of others? The answer is to be found in the special nature and function of the aid expected. India, in the first place, proposes to supply 80 percent of the capital; for her share in the effort India possesses the necessary resources in great plenty—in the form of unemployed geological wealth and underemployed manpower. The missing 20 percent is technology—in the form of skills, engineering tools, and plant equipment. These assets India does not yet possess in the self-regenerating abundance of the rich nations. Supplied from abroad, they will effect the junction of men and physical resources and produce 100 percent where the values are now zero or negative.

This is the nature of the aid India seeks from abroad. The function of this aid is equally decisive. In India's vision of development, aid is the offset to involuntary savings and the coercion necessary to secure them. Without aid in the volume projected, India would have to proceed, as other nations have done and are doing, to extract the last ounce of surplus from insufficiency by coercive deprivation.

External aid, in Indian planning, is the catalyst of development. It supplies the technology which brings manpower into reaction with resources at lower social pressures and temperatures. With external aid, it becomes possible to dream of carrying development forward without sacrifice of the living generation to the promised welfare of the next.

Aid, thus defined, is a misnomer. The word is as misleading, in view of the relation it implies between the poor and the rich nations, as the terms "particle" and "wave" in twentieth-century physics. The call for aid is not an appeal to the benevolence of the world; the condescension which burdens "aid" derives from the values of preindustrial technology and the days when some really had to go without. The demand for foreign aid (we are stuck with the term) which comes from India and other developing countries carries a rightful claim on the assets of twentieth-century civilization. The first among these assets is scientific knowledge and the power this gives man over nature. Science and technology are the heritage of all men, because they constitute the accumulative experience of our species. That experience came to sudden fruition in the West; but no exclusive title to it is vested in the West.

It is no coincidence that the popularization of well-being in the West has been followed by the popularization of citizenship. Now India proposes that the institutions of self-government shall not await economic development but shall develop concurrently. To this end it calls on the rich nations to help supply the needed technology.

Whatever the justice or propriety of this demand for foreign aid from the point of view of the poor countries, it exacts no sacrifice on the part of the rich. Science and technology are not diminished by the sharing of them. It is the information—the accumulative human experience—embodied in the tools and machines which the developing nations require, not the gross materials of which these artifacts of technology are made. The materials represent the least of the values and the smallest of the costs as well. The rich nations can respond to the demand for aid without perceptible cost to their well-being.

The rendering of external aid can, in fact, relieve the rich nations of peculiar temporary embarrassments which arise from the mismatching of the progress of their technology and the evolution of their social and economic institutions. What I shall have to say under this heading applies to a greater or lesser degree to all the industrialized countries. I shall call, however, on the example of the United States because I know it best and because it represents industrialization in its present most fully realized form. If my analysis leads into paradox from point to point, this flows from the no less incredible nature of the facts and figures of the American industrial economy. The logic of its abundance necessarily inverts values and habits of thought predicated on the more familiar experience of scarcity.

It cannot be said that foreign aid is a popular cause just now in American politics. India's needs—and the needs of other developing countries—have nevertheless found direct resonance and firm support in the second most powerful economic interest in our political system. This is American agriculture. The reason it responds so abundantly is that agriculture is in many ways the technically most progressive sector in the economy. Less than 7 percent of the American labor force is now engaged in agriculture. The contrast with the situation of India is drawn the more sharply in figures showing that most of the food consumed by the dwellers in American towns and cities comes from fewer than a million farms. More important to the food supply than the rest of the farmers and their hands are the 2 million workers in manufacturing who supply the fertilizers, tools, and machinery, and who process and pack the food. More important than farmers or workers are the 100,000, or so, agricultural technicians who keep the nation's agricultural productivity on a constantly steeper upward slope. Despite continuing reductions in the number of acres and man-hours, the American cornucopia continues to pour forth a greater flood of produce. The output at the farm is equivalent to 12,000 calories per day for each man, woman, and child in the land, enough to feed 1,000 million people. Americans feed some of it to animals, thereby exchanging carbohydrate for protein calories; we waste much of it; we give a great deal of it away; and still we have surpluses left over to keep in storage—50 to 150 percent of the annual requirement of all major grains.

Especially during the past fifteen years, these developments have confounded all efforts to manage American agriculture in accord with sound economic practices predicated on the assumption or the maintenance of scarcity. The American agriculturist continues to talk orthodox classical or scarcity economics on Sunday, as we say. But he has meanwhile learned to practice a sensible pragmatic kind of abundance economics on weekdays. Although he has not yet come to advocate anything so radical as production for use, he strongly favors production for production's sake.

Over the past decade one of the principal measures for sustaining the American agricultural economy against the crushing burden of its surpluses has consisted in shipping those surpluses overseas to feed the hungry. Since 1954, more than $9,000 million worth, 75 million tons, of agricultural commodities have been delivered to forty-four developing countries. Shipments of wheat alone have amounted to two entire bumper crops. India has been the principal recipient of these shipments—more than $2,000 million worth.

Under the legerdemain of Public Law 480, which sanctions this use of surpluses, the food is sold to the receiving government at world prices. The payment is taken, however, in nonconvertible currency and is loaned back to the recipient government to finance economic development programs. Sunday economics is thereby satisfied by the assurance that two dollars are made to grow in the place of one, and Public Law 480 stands secure on the American statute books.

In point of fact, as V. M. Dandekar has shown, food may be used to finance development. He found that food could be reckoned as 20 percent of the cost of an average infrastructure project in the Second Plan. Conversely, food may be said to exert a fivefold multiplier in the financing of development. Surely food is, as Edward S. Mason has said, "as good as gold"! Most important, this food has fed the hungry and has bought time wherever it has been shipped and consumed.

The benefits to the developing economy of the Public Law 480 food shipments are plain enough. One will scarcely credit, however, the benefits to the American economy. Here we have shipped out commodities of unquestionable intrinsic value; yet it is impossible to find that it cost Americans anything. The farmer was paid for his produce; in fact, as a result of these shipments and their salutary effect (salutary always means upward) on domestic American commodity prices (in accord with formulas written into the law and too complicated to go into here), the farmer received more than $1,000 million in extra income. The federal government, which unburdened its storage bins, made "savings in price support acquisitions, storage and interest" totaling $545 million in 1958 and 1959 on shipments which cost $668 million. Because the law reserves such shipping to American bottoms, the American merchant marine gained nearly $250 million in extra revenues from Public Law 480 during the first three years of its administration. Looking to the future, the farmers and their packers and shippers cheer themselves further with the prospect that these shipments have opened up new export markets for later dollar sales of American farm commodities.

Theoretically, the recipient countries have contracted to pay dollars someday for the foods and fibers shipped to them under Public Law 480; hopefully, they will be able to do so, in the long run, out of expanded national incomes. But the accounts have already been squared domestically inside the United States in the short run. The federal government set the wheel turning with its payments of the taxpayers' money to the farmer. These payments brought the farmer into the market as a customer for a long shopping list of consumers' and producers' goods the makers of which were glad to have a customer. The repercussions, multiplying the original transaction by two or three times, generated enough additional economic activity to bring income earners into tax brackets where they were liable for the additional taxes necessary to cover the original transaction.

Ultimately, of course, one can reckon up some real costs to the United States—some millions of tons of irreplaceable soluble minerals extracted from its topsoil. But this kind of cost awaits a more rational system of accounting to place it on the national ledger. Meanwhile, on the books kept by its scarcity accountants, the American economy shows nothing but plus signs.

If aid in the form of food works so well for both parties to the transaction, surely it is worthwhile to explore the possibilities of other aid in other forms. Professor P. C. Mahalanobis has set down some interesting reflections on the comparative advantages to the recipient country of various forms of aid-in-kind. He shows that the foreign exchange cost of the food grains needed to take care of the increase in India's population in the course of a Five-Year Plan—assuming population growth at the steady rate of 5 million per year—would come to Rs. 4,500 million. Alternatively, if India were to import the fertilizer needed to bring about the corresponding annual additions to the domestic production of food grains, the foreign exchange cost would be reduced to Rs. 1,350 million. Carrying the argument one step further, Mahalanobis shows that to import the fertilizer factories to make the fertilizer to grow the food grains would cost Rs. 120 million a year; this would bring the foreign exchange cost of nurturing the additions to the population during the Plan period down to Rs. 600 million.

In a parallel and independent set of calculations, Howard Cowden, an American businessman and student of public policy in agriculture, has arrived at a similar conclusion. Cowden observes that one dollar's worth of nitrogen applied where nitrogen is the limiting factor will get $4 in additional yield. An investment of $100 million in fertilizer factories will produce 3,000 tons of anhydrous ammonia a day; this much fertilizer will increase food production by $345 million a year. Cowden's logic encouraged the Cooperative League of the United States to urge that the U.S. Agency for International Development be authorized to join with its client countries in the building of $100 million worth of fertilizer factories, with Public Law 480 funds supplying the client's half of the necessary investment capital.

Since the logic works so well to this point, I, for one, am tempted to carry it still further, to the last step. My proposal is that India should set up the steel plant to make the steel to build the fertilizer factories to produce the fertilizer to grow the food grains. Going back to Mahalanobis's calculations, let us take steel to constitute a generous half of the foreign exchange cost of the fertilizer factories needed to keep food production increasing in step with population growth; the biggest cost, of course, is engineering. The equivalent steel plant would have to be imported only once; on going into production, it would produce the steel for a fertilizer factory each year thereafter. My proposal, therefore, calls for the ingot capacity needed to produce Rs. 60 million worth of finished steel. From India's own experience in building steel plants it appears that investment in ingot capacity runs about four times the value of its annual yield in finished steel and that foreign exchange nowadays necessarily represents 60 percent of such investment. The additional steelmaking capacity to which I am referring would thus cost Rs. 150 million in foreign exchange (Rs. 60 million × 4 × 0.60 = Rs. 144 million). By such investment India can reduce to one-thirtieth the Rs. 4,500 million foreign exchange expenditure for food computed in the first chapter of the Mahalanobis parable, an expenditure which would have risen to an annual rate of Rs. 1,500 million in the fifth year.

Of course, Rs. 150 million does not buy much in the way of a steel plant. India is a big country, however, and it can use a much bigger addition to its steel capacity. As Mahalanobis has observed: "In a big country it is possible and desirable to push back the manufacturing process to the utmost limit in order to expand continually its capacity to make investments increasingly out of its own domestic resources." A plant appropriate to the size of India is planned at Bokaro; at its full projected capacity of 4 million ingot tons, Bokaro will represent an investment of Rs. 7,000 million. The Rs. 60 million worth of steel needed to keep Indian agriculture supplied with new fertilizer plants will take up only 3.3 percent of its annual output.

Here is an insight into one of the paradoxes of industrial technology: food production increases in volume and efficiency precisely in ratio with its decline as a percentage of total economic activity. The United States operates the world's biggest agricultural establishment in terms of output, producing five times as much food as the people of the country need to sustain their nutrition; yet the agricultural sector accounts for less than 10 percent of the gross national product.

By this time, however, I do not suppose that anyone in India needs to be persuaded of the connection between agricultural productivity and industrialization. Rather it is the peoples of the rich nations who do not yet see the connection between food and steel. The Bokaro episode in the relations of the United States and India is a case in point. It is worth closer examination for what it reveals about the prospects for fruitful collaboration of the rich and the poor nations in the task of development.

Two apparent lines of argument were advanced against American assistance to the building of Bokaro; the two, in fact, come down to one. What hurt most was the technical critique which cited deficiencies in personnel, uncertainties as to raw materials, and inadequacies of transportation. Not long ago, in the United States, we heard the very same arguments advanced against Henry J. Kaiser's determination to build the first integrated steel plant in California—a plant which developed the lowest operating costs in the entire steel industry just as soon as Kaiser got it built. The supposedly technical critique of Bokaro may, therefore, be taken to be as purely ideological as the ideological arguments themselves. As for ideology, it must be admitted that Bokaro was the occasion for an old-fashioned camp meeting of Sunday economists. Listening to the clamor, one might find it hard to believe that the American businessman, factory worker, salesman, clerk, and engineer, no less than the American farmer, are learning to live and work in pragmatic accommodation with the rising tide of abundance which has swept away the premises of the official ideology.

Against all the precepts and injunctions of its formal economics, the American social order has been accepting entirely new values and priorities. From the ideological point of view, the most significant trend is the rising ratio of employment in the public as compared with the private sectors of the economy. More than half the new jobs created since 1950 are in the public sector. The functions served by the public sector do not yield what is ordinarily reckoned as "profit." The biggest expansion in jobs—more than a million new jobs in the decade—has come in teaching, a function which is characteristically public and which, in the United States, is the responsibility of local, municipal, and state budgets. In percentage terms, scientific research and engineering have been the most rapidly growing professions. Since the end of the Second World War the substantial growth—totaling perhaps 3 million jobs—in these elite functions has been financed almost entirely by the public sector.

The country has been compelled to these measures, in part, by the need to effect the distribution of its abundance via job-generated demand. Quite apart from the economic compulsions that justify such jobmaking, however, the values and the priorities have themselves begun to capture the public esteem. In this respect America lags behind other industrialized countries, where the public sector and its welfare functions have long turned over a larger portion of the gross national product and have held a higher moral priority. Nevertheless, the American solution to the management of abundance promises to set generous precedents for the world. Its public education system, for example, stands as the most truly democratic in design and motivation, even though the resolute effort to extend the best to all may slight both the brilliant and the backward.

If the United States had made more progress along these lines there would be no trouble about foreign aid. An enlightened and humane electorate would not fail to recognize its ethical obligations. But even now, in the present phase of America's transition from the want and scarcity which confine so much of mankind, there is support for the expectation that this country may lead the rest of the industrial nations to sponsor development on a meaningful scale.

The American economy faces a serious crisis. The rate of growth in recent years has not much exceeded the increase in population; a high percentage of the country's industrial plants remain chronically idle. Automatic factories, offices, and shops, in other words, have been producing as much poverty as wealth.

The true situation has been obscured by the hollow affluence of the war economy. Armaments, carrying as they do the absolute sanction of survival of the modern state, have had almost unquestioned command of the public treasury. The steady stream of funds pumped into this economic sink has directly and indirectly subsidized from at least 10 to as much as

20 percent of the nation's total economic activity. "Space," the "atom," and "big science" have held their catch basins under the overflow of "defense" and provided employment for many of the highly trained people that American society has been producing in such large numbers. It is a measure of the distance already traveled toward abundance that the economy can sponsor so much nonproductive and even purely wasteful activity not only without visible sacrifice but even as a covertly acknowledged means for maintaining consumer demand. The military budget has carried an additional ideological sanction in that it rediverts something more than 5 percent of the gross national product from the public directly back into the private sector.

The advance of technology has now, however, overtaken this makeshift arrangement. With the acquisition of overkill, armaments have lost their absolute claim on the treasury. Even in advance of a disarmament agreement a first small cut in the military budget is now before Congress.

Plainly the American economy must soon find other ways to sustain its activity at the present high rate. Since it will necessarily continue to produce surpluses beyond its own effective demand, the United States must invent new methods for disposing of them. Next to armaments the most convenient method would seem to be offered by foreign aid. As significant elements in the leadership of industry already realize, 80 cents out of each foreign aid dollar is spent within the borders of the United States. In the prospective American collaboration on the building of the Bokaro steel mill, for example, ". . . what has to be financed is rolling mills, presses, and other equipment that has to be imported into India to construct this mill. . . . Those things . . . cannot be bought with Indian rupees. They have to be bought—they will be bought entirely—in the United States if we finance this mill." Moreover, just as in the case of military expenditures, foreign aid takes the goods it buys out of the domestic market, thereby maintaining the scarcities that still keep the economic mechanism ticking.

Foreign aid thus calls on the same institutional relations between government and business as armaments and provides an equally direct channel for diverting funds from the public back into the private sector. A substantial foreign aid program would generate demand for the products of neglected and vital sectors of industry, including the heavy machinery builders and the machine-tool industry, the domestic business of which proceeds in cycles of "chickens today and feathers tomorrow." If shipments of fungible wheat can cultivate markets for American farmers, surely the installation of American machine tools would establish beachheads for future dollar markets. A radical expansion in foreign aid would also provide the most convenient way to soak up the surplus of engineering and research talent that is accumulating with the cutback in national defense and prestige expenditures. During the period of fifteen to twenty years which will be required, at a minimum, for the United States to bring its economic, social, and value systems into adjustment with the advent of automatic production, foreign aid can relieve many of the nation's internal stresses and strains. By the end of this period, given a sufficiently massive flow of aid, many of the developing countries, including India in particular, will have acquired the capacity for self-sustaining growth.

If foreign aid has not yet rallied the support of significant numbers of interested parties in the rich countries outside the American farm bloc, this is only because it has been conducted on such a pitifully inadequate scale. The rich nations variously inflate their claims as to the size of their foreign-aid programs. A dispassionate estimate is provided by the 1962 report on the economic and social consequences of disarmament to the Secretary General of the United Nations, prepared by an international group of expert consultants. The report indicates that the net flow of aid from rich to poor nations does not exceed $3,500–$4,000 million a year. This squares with independent estimates that the total rate of investment in the underdeveloped countries does not exceed $20,000 million—or less than 20 percent of the armament outlays of the rich nations. To the present flow of $3,500–$4,000 million, the United States contributes about 40 percent or something under $2,000 million, a figure which agrees well with the official American governmental figures, less the funds laid out for military purposes.

Since the American economy must soon find conveniently large open sluices for its surpluses—other than armaments—its foreign aid outlay could easily double. With disarmament, it might easily double again. If the United States were thus to take the lead in expanding the scale of external aid to the development of the poor countries, as it already leads in the rationing of the current trickle of aid, the total flow might equal or even exceed the $14,000 million figure projected by the committee of experts that considered this question for the General Assembly of the United Nations in 1951. With external aid on such a scale, the total investment programs of the poor nations could be boosted to as much as $100,000 million a year and might begin to approach the world's outlay for armaments. An increasing number of taxpayers in the United States and in other countries are ready to agree that foreign aid is a better buy.

There is another aspect to the relations of the rich and the poor nations, however, which presents considerable hazard to the generation of a significantly large flow of foreign aid. Alexander Herzen put the issue plainly:

So long as the educated minority, living off all previous generations, hardly guessed why life was so easy to live; so long as the majority, working day and night, did not quite realize why they received none of the fruits of their labor—both parties could believe this to be the natural order of things. . . . People often take prejudice or habit for truth and in that case feel no discomfort; but if they once realize that their truth is nonsense, the game is up. From then onwards it is only by force that a man can be compelled to do what he considers absurd.

The absurdity of the arrangements which enforce poverty in the twentieth century is becoming clear the world over. Development in some countries can come only with social and political as well as industrial revolution. In almost every country, it implies the revision or abrogation of the last thread of the colonial bonds that tie the developing country to its "home" country.

No matter what the leitmotif or the bloody detail of each

cycle as it gets under way, the extension of external aid can facilitate the underlying process of capital formation and soften its demands on the people. India's experience will prove decisive to the course of the development in all the other rising nations. For India's planners have plainly detailed the nature and the function of foreign aid and called for it on a scale sufficient to challenge the conscience as well as the interest of the peoples of the rich nations.

It is important to know that foreign aid on an adequate scale is technologically feasible and that it promises as much economic benefit to the rich as to the poor. The ground is cleared for confrontation of the moral issue. In the case of the United States, I am sure, external aid will begin to flow in significant volume just as soon as the Americans understand that their surpluses can lift the burden of history off the backs of the living generation.

Economic Development
by Democratic Planning

To Jawaharlal Nehru the most significant office he held in the government of India was chairman of its Planning Commission. Through that agency he articulated his vision of "economic development by democratic planning."

The Third Five-Year Plan, under way at the time of his death in May, 1964, closed out its books in 1965, a year short of the planned five years. Shortfalls in the industrialization plan, which depended so heavily on external assistance, were overtaken by the calamitous two-year drought of 1965–1966. By 1967, when Indian planners had recovered their nerve and were ready to launch the Fourth Five-Year Plan, the country had lost more than five years of economic growth that had been won at the painfully slow rate of 1.7 percent per year per person. The Fourth Plan places less reliance on external assistance and so sets less ambitious goals for industrial growth. With good monsoons, the "green revolution" has carried Indian agriculture to record harvests and brought in sight self-sufficiency in food supplies for the country.

Indira Gandhi succeeded her father not only as chairman of the Planning Commission but as a leader of the Indian civil servants and academic economists who are committed to her father's vision. One of her first acts, upon winning her own commanding majority in the parliament in 1971, was to reconstitute the Planning Commission as a government ministry. What will come of Indian planning and of Mrs. Gandhi's purpose, however, now hangs on the crisis in Pakistan.

There is a special irony in this turn of events. To the U.S. Department of State, its AID agency and their interested colleagues in the Department of Defense, the Moslem generals who run Pakistan have always made more sense than the Hindu intellectuals who govern India. Pakistan received economic assistance from the United States at twice the rate per capita. Since little of that aid went to East Pakistan, the rate of input to the benefit of the politically dominant western minority of the country ran even higher. The disintegration of the country and the murderous suppression of the East Pakistani political leadership following the first parliamentary election in the country's history show that the cold warriors of U.S. foreign policy have here, again, placed their bets unerringly upon the wrong horse.

The threat laid by these events to Indian security and stability is jugular. East Pakistan is East Bengal. West Bengal is a state in India. East Pakistan holds that country's largest Hindu population. The Bengali Hindus have been the principal victims of the repression from West Pakistan and number large among the 6 million who have fled across the border into India. Economic development in India must, therefore, take second or lesser priority to the choice of measures to keep West Bengal from hiving off with East Bengal to create Bangla Desh and to stay the Hindu communities of India from bloody reprisals against their Moslem fellow-villagers. These internal pressures may compel recognition by India of an East Pakistani government, followed by military intervention there.

In such event, our country's wrong bet in the Indian subcontinent will have set back the industrialization of India for a decade—and the vision of economic development by democratic planning, perhaps, forever.

The present essay was an attempt to tell the story to a business audience at Town Hall, Los Angeles, in June, 1964.

F LY east or west from the shores of America as far as you can go and you come to India. On the ground you will find you have traveled further in time than in space. India's multitudes live today variously as men have lived from pre-agricultural hunting and food-gathering times on into the squalor and convenience of the twentieth-century city. In most of the communities of India, you will find yourself in the world of 10,000 years ago: the Stone Age just behind and life secured by the technology of settled agriculture. As you will see, it is agricultural technology at an early stage: the mattock still does the work of the pick and shovel; the sickle, the work of the scythe.

Today in India, as 10,000 years ago, life is at risk to the uncertainties of production. In America, it is the vagaries of the process of distribution that make us insecure. In India, metabolic processes in the bodies of men and beasts supply the primary energy of the economy. In America, central power stations generate more than 6,000 kilowatt hours of electrical energy per person per year—the equivalent of 40 man-years of work done for each of us. Americans look forward to increasing material well-being, year after year. For Indians, from one day to the next, the question is survival: When do we eat?

America and India, so divided from each other in time and space, face each other across a still wider and deeper gap in human experience. They stand on opposite sides of the gulf between the rich and the poor. The American economy today is stumbling blindly into the unprecedented moral and social crises of abundance. The Indian economy strives against the ancient compulsions of scarcity.

Yet India and America—bewilderingly strange as they are to each other and so often strangers as well—are coming slowly to the understanding that they are embraced in an inescapable common destiny. It is not the "free world" that brings them together. India has never accepted election to this company which can count few practicing or even professing political democracies among its Asiatic, African, or Latin American members. Nor is it alone the moral stricture that declares no man is an island. The two nations are caught in a common fate because American policy in the next few years will critically affect the course of history in India and because the course of history in India in the immediate future will strongly determine whether our children will inherit an earth fit for human habitation—or whether the gulf between the rich and the poor is to be filled with the corpses of both.

The issue I am attempting to define here was put into world politics by Jawaharlal Nehru. From his earliest days in the Indian independence movement, Nehru recognized that the independence of his country could not be won until famine and epidemic disease stopped stalking the lives of his fellow countrymen. Nehru was not, of course, the first national leader in this century to call his newly liberated countrymen to industrial revolution. But Nehru had a grander objective: that India might carry through its industrial revolution without the violence and coercion and the pitiless use of human beings that have attended every industrial revolution in the past.

Nehru's vision has no historical precedent—surely not, as everyone will agree, in the contemporary industrial revolutions of China and Russia, and not in the classical industrial revolution of Britain and not in our own. Nehru was relying on two new developments in history. First: the technology of the industrial order is in being; it is ready "on the shelf" for export by the already industrialized nation and for import by the new industrializing nation. Second: the institutions of democracy bring the conscience of humanity into politics; it should be found desirable, therefore, as well as possible to carry out industrialization and the attendant radical changes in the social order by peaceful and noncoercive measures.

The true originality of Nehru's vision may be seen in the role he planned in it for our rich and successful democracy. This requires some preliminary explanation. As every businessman understands, it takes investment to increase production and it takes savings to put together the capital for investment. History shows, however, that most of the saving done by society is involuntary. Societies have devised a variety of appropriate measures of coercion to secure involuntary saving by its members, the severity of the measures being proportionate to the rate of saving. The examples of Russia and China gain ready assent for this proposition; it takes introspection of a somewhat painful kind to recognize its validity in our own history.

Rejecting coercion, Nehru and his colleagues projected a gradual increase in the rate of saving by the Indian economy from a starting level of 5 percent of national income upward over the course of fifteen or twenty years, especially as industrialization proceeded, to a high of 20 to 25 percent. Plainly, the rate of savings that could be secured voluntarily in the early years of India's industrialization would be inadequate to sustain significant development. This is where America was to play its role. Aid from America and other countries that could afford it would fill the gap until growth in India became self-sustaining. The kind of aid projected was the kind we could most easily afford: aid-in-kind, the surpluses of our farms and factories that do not find markets at home or elsewhere.

It is a fantastic scheme and, from some points of view, loaded with effrontery. Yet the scheme has worked or shown, at least, that it can be made to work. After seventeen years, India remains a parliamentary democracy, one of a select few underdeveloped countries not headed by a military dictator. The Indian economy has made encouraging progress toward industrialization. Moreover, it has received aid in the form projected from America and other nations. Progress has been agonizingly slow, and growth has fallen short of the objectives set by India's planners. But the fault is not all with the Indians and their planning. It happens also that the flow of aid from our shores has fallen short of expectation. Significantly, the record shows that growth has fallen short of plan roughly in proportion to the shortfall in aid. Meanwhile population increase has offset much of the gain in production.

To recover lost ground and get back on schedule, India's economic growth must accelerate. If that acceleration is to be accomplished without jettisoning the nation's constitution, then aid must flow into India at a correspondingly accelerating rate. It is plain that the future of India depends upon the success with which U.S. citizens play the role assigned to them. Nehru is no longer alive to personify his plan and to cajole and extort our help. Americans must therefore summon the necessary understanding and motivation on their own. We must come to know India better.

Huge numbers describe India. Her 450 million people, more than twice the U.S. population, occupy a subcontinent of 1,265,000 square miles, about one-third the U.S. land area. The territory reaches from the world's loftiest highlands in the Himalayas to the oceanic tropics and on the east–west line from China to Arabia. One-fifth of the land is covered by forest; two-fifths is under cultivation. Of the 325 million acres under cultivation, 70 million are irrigated, and there is water in abundance to irrigate 100 million more. India's mineral resources include one of the world's major reserves of iron—20 billion tons of high-grade ore—an abundance of bauxite and manganese, and important deposits of such exotic minerals as gold and mica. Her energy resources begin with 50 billion tons of coal including 2 billion tons of coking grade; 40 million potential kilowatts of falling water; limitless horizons of monazite sands containing the nuclear fuel thorium, and 300 days per year of dazzling sunlight which may soon be made to yield more "atomic" energy than the thorium sands.

It is not overpopulation, on the one hand, or lack of resources, on the other, that places India among the poor nations. What India lacks is the catalyst of industrial technology that makes resources serve human need and demand. Even in technology, however, India has a long lead on other underdeveloped nations; for India has a highly cultivated and educated leadership, including essential cadres of scientists and engineers, who have already demonstrated their ability to manage the initial phases of industrial revolution.

At this point it is well to urge the caution that there is more than one India. The country's constitution recognizes thirteen languages, and the speakers of some would rather accept English as the country's lingua franca than submit to instruction in

Hindi, which is the language of no more than the largest minority. For purposes of the present discussion, we shall think of two Indias. The first India I want to evoke here does not fit so easily into our preconceptions. On the spot as well as in the mind's eye, it dissolves constantly into its setting in the other India from which I am attempting to abstract it. This first India is an urban, literate, mercantile society of about 50 million people. If you could credit the country's entire GNP to their account—as you well might, because the Indian monetary economy is so largely theirs—then you would be dealing with an economy that stands comparison with, say, Brazil or even Mexico. In fact, with its denominators so reduced on paper, modern India comes out ahead of either Latin American country on most indexes of the world economy, including the industrial.

The denominators are not so easily written down, however, because they are people. These are the people of the other India. They are village-agricultural folk, in the main, but they squat in the shanty suburbs of modern India, serve as its dray animals, and die on its sidewalks, so their insistent, multitudinous presence can nowhere be escaped.

A great many of the proprietors, managers, technicians, politicians, and other operators of modern India like their country the way it is. They would settle for India's present relation to the world economy and would, in particular, enjoy the exchange of their country's exports for the sorts of things their country does not make and they personally are in the market to buy. While they are imbued with enterprise to go with their appetites, they have a high, old-fashioned preference for capital liquidity; they approach fixed investment in infrastructure and heavy industry with a short-range, wary eye. From their point of vantage, they see no reason why industrialization should be revolutionary. The imports of this economy—a shopping list of consumer durable goods and manufacturers' current production materials and components—just about balances the country's exports.

The face of modern India more familiar to Americans is that of its political leadership dominated by the patrician arrogance of Nehru. What the world outside does not appreciate is that this political leadership, which carried the revolution against the British crown, is now reduced to opposition in the country it has liberated. The revolution Nehru and his colleagues seek to foment is industrial. It turns out that they must press this revolution against the inertia of the native Indian ownership that has succeeded the British in possession of economic power in the country. In keeping with the Gandhi tradition of non-violence and the Nehru commitment to noncoercion, their revolutionary weapon is the plan.

The planners of India's transition from agricultural to industrial civilization must reckon with elements too subtle to be counted. They cannot borrow the models of Western economics that so conveniently treat people as a collection of identically motivated economic men responsive to simple forces of supply and demand. The Indian planner must find his models in biology—in the ordered complexity of the living cell and the orchestration of highly differentiated cells in the function of the tissue.

Some 80 percent of the population—360 million people—live on the land huddled in the mud huts of 500,000 villages. But the village is not the ultimate unit of social and economic life. The 1,000 or so inhabitants of the average-sized village may be segregated in as many as a score of castes that tie their members more closely to kinsmen in neighboring and distant villages and may even set them in adversary relationship to their fellow villagers.

The intricacies of caste defy the understanding of the most learned students of the subject. Here it is enough to observe that the castes divide India's population into no fewer than 12,000 endogamous groups. Within these groups intramarriage has persisted so long that they are distinguishable by significant variations in blood type and other genetic markers. More easily recognizable differences in religious belief and observance, dietary customs, kinship systems, clothing, trades and skills, and even languages perpetuate caste in the society of India today. Such social tissue is not subject to easy manipulation. It shows high capacity to absorb and dissipate energy and to frustrate initiative applied from the center or from the outside.

Resistant as the caste communities have proved to change

and divisively as they have worked against the emergence of India's national identity, they have a place in most Indian Utopias, whether the design comes from the left or the right. It is not the aim of India's leaders to bring their country into the same kind of paved-over, rubber-tired, smog-shrouded Utopia we have built here in America. The village and the caste-community may function in ways that cannot now be predicted to cherish the personal sanctity of the individual and to place him in meaningful relationship to society as a whole. With such objectives in view, the Indians are striving, through change in law and custom, to rid the caste system of its evils. Significant gains have been made in the democratization of the intercaste hierarchy and in lifting the curse of untouchability from the outcastes. From India, the rest of the world may yet learn how to make the industrial system a more humane environment.

The first seventeen years of independence have seen India's central political leadership begin to bring the country together into a nation and launch its industrial revolution. The figures are impressive and call for revision of the American folk image of the Hindu as a bemused mystic. Agricultural output increased 37 percent and industrial output increased 100 percent during the first decade of planned growth from 1951 to 1961. The output of steel doubled, from 1.5 million to 3.4 million ingot tons; the output of electrical power climbed from less than 3 million kilowatt hours to more than 6 million. The percentage of young people in secondary schools and colleges multiplied two and a half times, and the number of children in elementary schools was pushed upward to 60 percent of all children of school age.

India is so vast, however, and its numbers so astronomical that the bravest gains are overwhelmed by sheer magnitude. The 40 percent of the children for whom teachers and schools must still be found, for example, number 23 million; that is as many children of the same age as America now sends to school. During the first decade of planning, the nation's population increased by 80 million, reducing gain in total output to a 19 percent gain in output *per caput*. The bottom 10 percent on the income scale have as little as a nickel a day to spend on all their needs; the consumption of the richest 5 percent aggregates no more than 50 cents a day.

In India, even despair is a luxury. Indians who know these statistics firsthand in the lives of their kinsmen and fellow citizens manage to carry on with hope and determination. Their morale depends strongly on the disposition of the industrial nations to join in their task.

In fact, it can be argued that Indian planning is scaled from the planners' estimate of available foreign aid. In India's overall planning, 80 percent of the necessary capital is represented by labor and construction materials. These India can readily supply from her abundance of human resources and geological wealth. The 20 percent that India cannot so readily supply is technology. It is technology, however, that puts men and resources to work. Presumably India might buy it. But the world-wide decline in commodity prices since the end of World War II and rising tariff barriers around the markets of the industrial nations have given India and most other developing countries a negative foreign trade balance. India must therefore seek technology in the form of aid. The aid must come if India is to proceed with the task of democratic, noncoerced development. The absolute amount of aid available at any time thus sets the size of the 20 percent which, multiplied by five, sets the size of the 100 percent of the total investment effort.

The figures for the present period are illuminating. The Third Five-Year Plan has as its goal 100 billion rupees of investment effort. That figure is associated with a 25 billion-rupee foreign exchange deficit for the five-year period. By no coincidence at all the plan calls for the importing of 25 billion rupees' worth of machinery and equipment, components, and intermediate products for raising the productive capacity of the country. The deficit represented by capital goods equals the aid from abroad, in grants and long-term loans, on which the whole five-year plan turns. With the country's foreign exchange otherwise in balance, India's industrial growth thus depends upon how much capital assistance its political leaders and planners can cadge from abroad.

Let me assure you, incidentally, that there is no problem

about "absorption" here. Absorption is one of the cant words of the new overnight experts in foreign aid who have turned up to administer the programs in Washington and write about them in learned journals. It refers to the capacity of the receiving country to absorb capital inputs. To say a country can absorb just so much—by reason of its limitations in resources, trained manpower and, the ultimate condescension, managerial talent —is to assure the donor country that it has done its bit and the donee that it has got all it should get. The most satisfied party to this transaction is the expert in the new economics of development; he has helped, in his way, to make this the best of all possible worlds.

For very small and very new sovereignties, such as those emerging in the Caribbean, there may be a limit to the rate of capital investment in resort hotels, for example, that they can absorb. It would be difficult to find, however, a country that has been bowled over by any excess of capital and technical assistance. On the contrary, countries that have been the object of the military-strategic anxiety of the United States, such as South Korea and Formosa, have been supplied with assistance above and beyond the nice calculations of theories of development; they are now racing into self-sustained growth.

India belongs to neither of these categories. It is a big country with an ancient culture, abundant resources, and a leadership committed to protecting its independence. The 25 billion-rupee foreign exchange deficit projected in the Third Five-Year Plan comes not from the econometrics of development but from the political calculus that says this is the most India can hope to get by way of foreign aid on acceptable terms. India could, in fact, absorb capital inputs at two and three times that rate in a much more ambitious and fruitful Third Five-Year Plan.

Translated into dollars, this means that India seeks a mere $5 billion of aid from abroad over the entire plan period. The prospects of India could be doubled by an additional $5 billion and tripled by an additional $10 billion of foreign aid. If the United States were to continue contributing 40 percent of the total, the biggest of these figures would call on the U.S. taxpayer for just $100 million per month for the next five years.

Let us consider these outlays from the point of view of the United States taxpayer. In the first place, starting with the Marshall Plan, experience indicates that at least $80 million a month would be spent inside this country on its own goods. Something even closer to $100 million would stay home in the case of the capital goods India is looking to us for. The customer for these goods, so far as American industry is concerned, is our own federal government. Industry is used to doing business with the government by now, so there is nothing novel about this transaction. It happens that Los Angeles, with its military and space industries, would not find much increase in the business it does with its biggest customer. But, surely, Los Angeles will not begrudge Pittsburgh, the Naugatuck Valley, and other centers of our more senior technologies a mere $100 million per month of the federal government's purchase orders. For the taxpayer, these purchases mean not only the direct employment of the makers of the capital goods—and so more fellow taxpayers—but the employment of additional taxpayers, perhaps including himself, to make consumer goods for purchase by the paychecks of the makers of the capital goods. In other words, a dollar spent for foreign aid has the same expansive effect on the domestic economy as military expenditure. American taxpayers have learned to find comfort in this legerdemain over the past twenty-five years.

To this taxpayer, investment in economic development looks like a better buy. Suppose as much as 20 cents out of each dollar is spent abroad. That 20 cents excites a whole dollar's worth of economic activity here at home in the industries that produce the capital goods on India's shopping list. Upon arrival at its destination, that dollar's worth of goods excites $5 worth of economic development. In its total net effect, therefore, the original 20 cents brings a 25-fold return. But the 25-fold economic return does not begin to measure what can be bought here for the future of both the developed and the developing countries and, in particular, for the safety and sanity of the children of America and India.

Second Thoughts
on Birth Control

THE students of Samuel Z. Levine, chairman emeritus of the Department of Pediatrics of Cornell Medical College, maintain a fund in his honor which he had the pleasure of spending until his death in July, 1971. He devoted the money to the adaptation of modern medicine to the circumstances of the poor countries with the special object of coping with the infections in children that account for most of the deaths in the high death rates of those countries. This essay was prepared for the annual dinner of the Samuel Z. Levine Foundation in March, 1965.

T HE world population has doubled twice during the last 300 years. The first doubling took a little more than 200 years; the second, somewhat less than a century. The rate of increase continues to increase. Today, the population is gaining a nation the size of the United Kingdom, 50 million people, every year; a city of nearly 150,000 people every day;

more than 6,000 per hour; 10 per second. At this rate, the population will have doubled again before the end of this century.

These are the numbers that conjure up the "population explosion." No special talent for arithmetic is required to extrapolate the trend to the absurdity of "standing room only." Far short of absurdity, checks on the growth of population must come inexorably into play. Thomas Malthus, early in the nineteenth century, set out his famous calculation of the divergent growth of the population and its means of subsistence: ". . . supposing the present population equal to 1,000 millions, the human species would increase as the numbers 1, 2, 4, 8, 16, 32, 64, 128, 256, and subsistence as 1, 2, 3, 4, 5, 6, 7, 8, 9, in two centuries, the population would be to the means of subsistence as 256 to 9. . . ." Long before arrival at this absurdity "vice and misery" would take their toll unless men learned to practice "moral restraint."

If the population increase has not followed the geometric course projected by Malthus in the nearly two centuries since he wrote, this may be credited in part to the positive checks arising from vice and misery: "all unwholesome occupations, severe labor and exposure to the seasons, extreme poverty, bad nursing of children, great towns, excesses of all kinds, the whole train of common diseases, wars, plagues, and famine." Now, with the next geometric milestone looming before the end of this century, men of good will everywhere look for humane methods to brake the growth of population by rational choice. Techniques of contraception are available to reduce the stringencies of moral restraint. In America, the planned-parenthood movement, founded by Margaret Sanger, has promoted the practice of birth control widely in the population and reduced the barriers of prudery that stood against public education and public assistance in the practice. The same leadership, turning to concern for world population growth, has recently succeeded in establishing the supply of contraceptives and instruction in their use as a major feature of the technical assistance offered by the United States government to the new, poor countries of the former colonial continents.

It is the latter development that prompts these second thoughts on birth control. Apart from urging reconsideration of the policy on grounds of the uncertain practicability of present contraceptive techniques in the village culture of these preindustrial societies and the affront ("too many Indians") felt by so many of the people to whom this kind of help is proffered by the richest nation in the world, this discourse will set out an entirely different strategy for world population control.

In the first place, no agency short of the thermonuclear apocalypse can prevent the next doubling of the world's population, to which the growth of population in the developing countries will contribute the overwhelming plurality of numbers. It will be argued that the right strategy is to bring on the population explosion in those countries as rapidly and as early as possible. To this end the rich countries should speed up the industrialization of the poor countries with generous inputs of economic and technical assistance. The object of this strategy is to give the next generation and the next a smaller total population base from which they may seek an optimum accommodation of the species to the finite resources of the planet. Industrial revolution, in other words, offers a more direct route to population control than contraception. In the special circumstances of the poor countries, it will be shown, pediatrics has more to offer in the immediate future than gynecology. As for those well-meaning, single-minded advocates of birth control, they can do more for the cause in the developing countries not by pushing their propaganda at the natives but by encouraging fundamental research in human reproduction here at home.

Some comfort may be gained from the thought that the present population explosion is not the first but the third in the history of mankind. The chart that Edward S. Deevey designed to make this point (see graph, opposite) spreads out the three episodes in logarithmic scale. Though the chart thus understates the visual impact of the huge recent increases, it puts those numbers in perspective. For the chart brings out the connection between human population and the technology at its command. The three explosions each attend a revolution in technology. The present explosion is not out of scale with contemporary technology or, manifestly, it would not be occurring. Moreover, the abundance afforded by industrial technology has begun to displace vice and misery as the primary check on population growth among people who have the good fortune to enjoy it.

The first revolution in technology got under way more than a million years ago. The term "revolution" is used here somewhat in its geologic sense; the revolution proceeded, in fact, on the time scale of biologic evolution. This was the toolmaking revolution, and it came along in step with the evolution of the human species, at once a cause and an effect of the evolution of man. Just when man came on the scene cannot be dated with certainty. According to Loren Eiseley, speaking on one of the few occasions when a scientist has addressed the American Academy of Arts and Letters, man had better not try to set such a date:

All along the evolutionary road it could have been said "This is man," if there had been such a magical delineating and self-freezing word. It could have immobilized us at any step of our journey. It could have held us hanging to the bough from which we

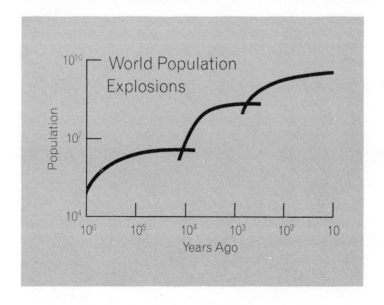

actually dropped; it could have kept us cowering, small-brained and helpless, whenever the great cats came through the reeds. It could have stricken us with terror before the fire that was later to be our warmth and weapon against Ice Age cold. At any step of the way, the word man, in retrospect, could be said to have meant just such final limits.

Along that road, emergent man recorded his progress in the stone tools that Deevey has called "the most common fossil of the Pleistocene." From the abundance of tools, he estimates that the number of people "ever born" during this time may have approached 70 billion. Few survived infancy, and few survived to reproduce themselves more than once or twice. The generations succeeded one another rapidly. From the extravagant proliferation of lives, natural selection chose *Homo sapiens*. By about 100,000 years ago, the success of this first revolution had brought the population up to 5 million.

The second revolution proceeded at a faster pace. Beginning perhaps 10,000 years ago and within a few millennia, the elaboration of agricultural technology raised the capacity of the earth to sustain human population by two orders of magnitude, to one person per square kilometer, and the population soared to more than 100 million. Progress in technology was slow; a man did not see the productivity of his labor increase during his lifetime. The geometric pressure of human fertility was relieved, from one generation to the next, by increase in the acreage of soil under cultivation. Most of the people most of the time lived on intimate terms with misery.

History kept scanty demographic records in this period. Some idea of what life was like can be gained from the condition of man living today in India. More than 80 percent of the people there are bound to the soil, scratching out an insecure subsistence and rendering up a meager surplus for export to the urban economy. Until recently life expectancy had not increased much above that of preagricultural man. A woman who could endure the experience might undergo six to eight pregnancies. Of her offspring, she might expect to see half die before the fifth day and, of the survivors, half before the fifth year.

For the small percentage of the population that lived off the rest, life was a happier story. In particular, they could expect to see their children survive those first five days and first five years. Prudent and frugal families sought to limit the number of their offspring. It is well known that the upper classes practiced birth control, whether by moral restraint, by techniques for avoiding conception known from earliest times, by abortion, or by the draconian method of infanticide. Cornelia's jewels—the Gracchi, Tiberius and Gaius—were two.

Most people today live in countries with a high death rate and a high birth rate. This gives a chart of their numbers by age groups the profile of a squat triangle (see graph, p. 218). At the bottom, in the broad base, are the infants and children below a median age, for the whole population, in the early teens; a narrowing steeple of older age groups reaches above. The success of the agricultural-urban revolution is recorded in the growth of the world population from about 100 million 2,000 years ago to 1,000 million in Malthus's time.

Malthus propounded his principle of population just as the third population explosion was gaining momentum. In middle-class disdain he recoiled from the degraded condition of Britain's urban proletariat. He did not recognize in their poverty the process of capital formation by involuntary saving that was financing the third revolution in technology. Malthus had more compassion and hope for his fellow-men, however, than some neo-Malthusians of our day.

With this revolution, technological change accelerated to a pace that must now be clocked in decades and half-decades. The productivity of human labor, amplified by inanimate energy and by understanding of nature, broke through the arithmetic constraint on increase in production.

The success of the industrial revolution is recorded in the spectacular expansion of the European population. From 100 million, 15 percent of the total world population at the beginning of the eighteenth century, the Europeans in Europe and overseas multiplied to 700 million and about 25 percent of the world population 250 years later. They occupied the continents of America and Australia, where primitive cultures could not stand in their way, and, similarly, the congenial portions of

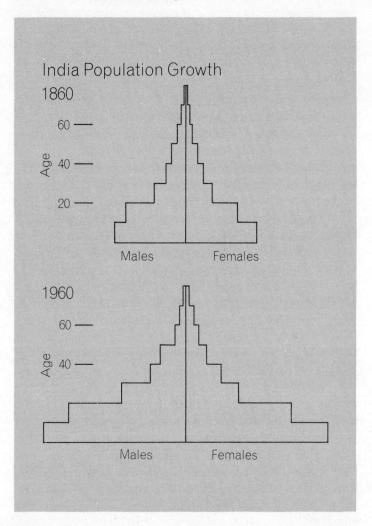

India Population Growth

1860

Age

60 —
40 —
20 —

Males   Females

1960

Age

60 —
40 —

Males   Females

Africa. They have now brought in sight the first world civilization.

The European population explosion reflects a historically unprecedented decline in the death rate; the decline shows up in the population of one nation of Europeans after the other, starting in the eighteenth century, and continues into the present. The explosion would have carried their numbers still higher, however, if the decline in their death rates had not been accompanied by an equally unprecedented decline in their birth rates. Thus, from 1800 to 1960, the death rate in the Scandinavian countries fell from 25 per 1,000 to less than 10; the birth rate came down in parallel from 32 per 1,000 to 16.

The trend in both death and birth rates follows, of course, from the popularization of material well-being. In country after country, increase in the supply of material goods proceeded to outrun increase in the population. Incomes per capita rose as the abundance trickled down even to the most disadvantaged members of each national household. Entire populations came to enjoy a standard of assured nutrition, living space, and material comfort formerly reserved for the fortunate minority at the top of society. People in these circumstances not only lived longer, they also made the simple calculation which shows that, with fewer mouths to feed, there can be more for those already born. It is notable that half or more of the decline in the birth rate in Scandinavia antedates the arrival of aesthetically sophisticated techniques of contraception.

Today the industrial countries, especially those where incomes exceed $1,000 *per caput,* present an entirely new demographic profile. The sides of the triangle, that is the steps from each age group to the next older, go up more steeply. Compared to the classical profile, relatively larger numbers of persons appear in each age group above the bottom (see graph, p. 220). It is this filling in and evening out of age groups that account for almost the entire doubling of the U.S. population, from 75 million to 150 million, over the half-century from 1900 to 1950. The increase in the numbers in the age groups above ten years of age comes to more than 60 million. Significantly, the size of the age groups below ten years of age remained constant and even declined over the years up to 1940 and increased only with the postwar baby boom. That boom now appears to have represented carry-back and carry-forward of declines in the birth rate that preceded and followed it. The effect of even a small increase in the width of the base is amplified, however, by the high rate at which individuals survive into the age groups above.

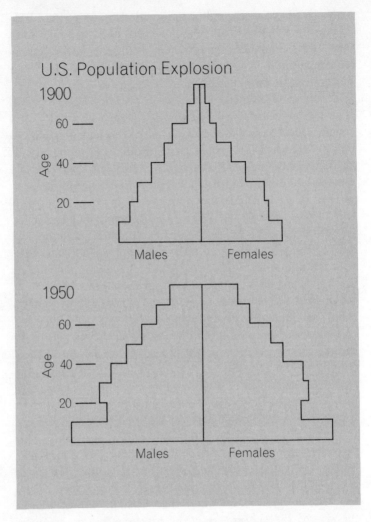

U.S. Population Explosion
1900

1950

The same profile characterizes the populations of other industrial, low death rate–low birth rate countries. It would
appear, therefore, that the population explosion of the European peoples has spent its momentum. That the individual,
personal impulses and decisions reflected in such change in the
growth and structure of populations are not peculiar to Europeans is shown by the later, rapid transformation of the Japanese demographic profile. The first nation of Asia to join the

circle of "$1,000 countries," Japan now sets the world the example of substantially constant population size. By contrast, the United States sets perhaps the worst example, with a 25 percent population increase since 1950, at an average rate of nearly 1.7 percent and with substantially no immigration. In the United States and most other industrial countries birth rates continue to exceed death rates. Population trends remain the summation of personal decisions motivated by private concerns. Populations have not yet begun to limit their fertility in the interests of the next generation of the species.

A population explosion of the same new historic kind impends, or has started, among the peoples of the "$100 countries." Over the past century their populations have been growing, but at the old historic pace, held in check by insufficiency. The population of India grew, for example, from 250 million to 450 million people over the past century. This was at less than half the rate of the concurrent U.S. population growth. The necessary increase in production was achieved by bringing more land under cultivation, and the extension of agriculture to poorer soils was accompanied by a sharpening of the endemic misery. Over most of the century, there was little change in the classical triangular shape of the country's demographic profile. The base of the triangle, measuring the numbers in the younger age groups, grew wider (see graph, p. 215).

Population growth in these countries has been quickening since the war, however, owing to abrupt reduction in their death rates. This has been accomplished by the most portable of the Western technologies, that of public health. With the eradication of malaria in Ceylon, by the spraying of DDT, the death rate dropped from 22 to 12 per 1,000 in seven years. In India the death rate has fallen since the war from 27 to 19 per 1,000. The measures that brought these changes affect principally the survival of those over five days old. For the newborn the odds remain at 50–50; tetanus inoculated by the midwife's knife kills 25 percent. For the survivors the prospect continues to be, in the words of Maurice Pate of the United Nations International Children's Emergency Fund, "a short life, a sick life, a hungry life."

The reduction in their death rates has now started these

people into the explosive population growth that comes with change in the demographic profile. Survivors are populating the age groups above the youngest. The question for history in the last decades of this century is: How fast can the poor nations escape the Malthusian cul-de-sac? The speed with which they make their escape will determine the ultimate size to which their populations must grow. If they could, as is not likely, match the economic growth rate of the United States from 1900 to 1950—if geometric population increase could be answered by geometric increase in production—then their population growth could be contained at about 100 percent of their present size. In that case, the growth would come from filling out of the age groups in the demographic profile above a base line held at its present width. If economic growth proceeds more slowly—if the geometric population pressure continues to be accommodated by not much more than arithmetic increase in production—then the base of the triangle must continue to grow wider, as it has over the last century. That would mean increase in the size of the population on hand to be doubled in the ultimate escape from the Malthusian cul-de-sac—if that escape is ever to be made.

The strategy for world population control commended by this analysis is to hasten the economic development of the $100 countries. This strategy assumes, of course, that the percolation of material well-being among these people will establish the motivation for fertility control among them as it did among the Europeans and now the Japanese. There is no reason to assume otherwise. What is more, there is evidence to encourage this assumption.

For the Indians, as for other village-agricultural people, surviving children are insurance against the infirmities of age that come to most of them before they are old. They have had to maintain their fertility rate high against their high death rate. On the other hand, once babies begin surviving, their hungry mouths must excite parental concern of another kind. Figures to prove this surmise in the case of the Indians are not automatically available. India is a big and poor country with most of its population concealed in half a million villages, many of them never visited by a census taker. Sample studies conducted for the government by the Indian Statistical Institute at Calcutta, however, provide important insights into the dynamics of India's present population growth.

These studies show the same relationship of family size to household expenditures in all parts of the country. The most wretched have no surviving children. In households above the bottom and up to a "critical" expenditure, the number of children increases with expenditure. Then, above that critical expenditure, the average number of children per household decreases (see graph, below).

The impressive and touching fact is that the critical expenditure is close to the average expenditure in each region. The discouraging fact is that the average expenditure is everywhere high above the median. Unequal distribution of property and incomes holds the great mass of India's families below the critical hump in the curve at which family size goes into inverse relationship to further increase in household expenditures. In consequence, and thanks to death control, India's population growth is now running above 2 percent, more than three times the average rate of the last century.

This suggests a refinement on the strategy of population con-

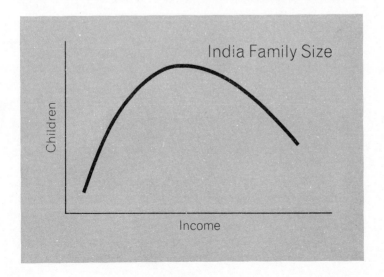

trol by economic growth. If increase in gross national product continues to be distributed to the population in accord with the prevailing inequities in incomes, it will require a very huge increase—and, at the present rate of development, a very long time—to set the stage for reduction in the fertility rate. Distribution of economic gains to the lower-income groups, on the other hand, would push more families more rapidly up to and over the hump on the family-size curve. Increase in incomes on the average will not promote population control so efficiently as increase in the incomes of families below the median. This is one of the motifs in "the socialist pattern of society."

The strategy here urged and its refinement accord well with Thomas Malthus's own prescription for moderating the severity of his principle of population. He recognized that death control is the first requisite: "Every loss of a child from the consequences of poverty must evidently be preceded and accompanied by great misery to individuals and . . . loss to the nation. . . . Consequently, in every point of view, a decrease of mortality at all ages is what we ought to aim at." He saw that a substantial population increase would have to be absorbed: "We cannot however effect this object, without first crowding the population in some degree by making more children grow up to manhood." The argument now showed: "It is the diffusion of luxury therefore among the mass of the people, and not an excess of it in a few, that seems to be most advantageous, both with regard to national wealth and national happiness . . . [O]ur best-grounded expectations . . . are founded in the prospect of an increase in the relative proportions of the middle parts."

Malthus concludes this passage with a prayer:

[W]e might even venture to indulge our hope that at some future period the processes for abridging human labor, the progress of which has of late years been so rapid, might ultimately supply all the wants of the most wealthy society with less personal effort than at present; and if they did not diminish the severity of individual exertion, might, at least, diminish the number of those employed in severe toil. If the lowest classes of society were thus diminished, and the middle classes increased, each laborer might indulge a more rational hope of rising by diligence and exertion into better station; the rewards of industry and virtue would be increased in number; the lottery of human society would appear to consist of fewer blanks and more prizes. . . .

He did not anticipate that such "rational hope" would motivate individuals to inhibit their fertility. The most Malthus expected was that moral restraint might move "the laborer" to postpone marriage until he could provide for his six or more children.

The moral restraint most widely practiced in Indian households that are secure in the survival of their children is abstinence. This conclusion is suggested, in any case, by the not yet published results of a large-scale controlled experiment in the promotion of the use of contraceptives. It was conducted under the auspices of an American philanthropic foundation in the villages served by the village medical center at Najafghar, outside Delhi. After four or five years, the population that had been sold on the use of contraceptives showed a measurably higher fertility rate than the matched population living in innocence.

Evidently, fertility control is a question of motivation before it becomes an object for exercise of technological ingenuity. For such ingenuity as is available, the more useful service would be to raise the level of motivation. Pediatrics can provide the decisive motivation bodied forth in surviving babies. Working parties fielded in the Philippines by the Samuel Z. Levine Foundation have shown that villagers can be trained to pull infants through the crisis of dehydration brought on by the infant diarrheas and that the capital equipment required can be brought within their means and competence. Though a dung fire does not give off BTU's at a high rate, it ought to be possible to sterilize that midwife's knife. By elimination of tetanus alone, this would reduce the death rate among newborns in India by 25 percent.

Good health promotes survival, of course, even more effectively than medicine. One must accept the judgment of Stephen Plank, of the Harvard School of Public Health: "The conditions required for the adoption of modern agricultural techniques are largely the same as those that would tend to lower birth rates, and vice versa." Modern agricultural techniques come embedded in the apparatus and culture of industrial civilization. They call for assured water supplies, biochemistry, genetics, tools, and machinery on the input side and storage, processing, and transportation on the output side. Nothing less than industrial revolution can create the conditions required for population control. Developing countries now heading into humanity's third population explosion ask the rich industrial nations for contributions more significant than the gadgetry of contraception.

This is not to say that simpler and surer contraceptive devices and procedures might not promote fertility control in primitive circumstances at lower levels of motivation. Quite apart from the results of Najafghar, however, it is apparent that the present technology has little to offer villagers who lack privacy, running water, and medical supervision. The energy and good will available for the promotion of birth control abroad ought to be redirected to the encouragement and support of fundamental research. After a survey of the work now going on, Solly Zuckerman has concluded that the sights could be considerably raised:

This method of isolated observation illuminates only a small part of the picture and distorts the whole. At every turn the process of conception can be shown to be a sequential, coordinated and overlapping series of mechanisms which seem to overinsure against any possibility of physiological breakdown. Few, if any, of the steps in the process seem to be mechanisms in which single factors are concerned. Equally, when one comes to consider the general problem of fertility, one sees that its overt and quantitative expression is the resultant of many different factors or parameters which exercise their various influences at every step in the process of reproduction.

Zuckerman is saying, in effect, that the human species would not be here if its reproductive physiology did not stubbornly resist interference. The research effort appropriate to the inherent interest and vital importance of this topic has yet to be launched.

Workers in the birth-control movement in the United States cannot find a more appropriate audience for their propaganda than their fellow citizens. The birth rate of this country, at 21 per 1,000, profligately exceeds the death rate at 9 or 10 per 1,000. Public education in these matters might usefully begin with the official economists and the oracles of the financial pages. Their message of perpetual economic growth counts heavily upon sustained increase in the number of new households from decade to decade. Domestic economic growth and containment of the world population explosion both demand the same capital remedy: massive investment and technical and economic assistance by the United States to hasten the industrial revolution of the developing countries.

A World
Free of Want

The following is a memorandum I contributed in March, 1966, to the deliberations of a discussion group on "science and the national interest."

WE HAVE been considering the implications for the conduct of U.S. foreign policy—and for the definition of the national interest itself—of the contemporary revolution in the technology of war. We turn now to examine what we should all prefer to believe is the obverse of the same coin. Our task is to consider the implications for national policy of the constructive potential of science and technology, especially with respect to the economic development of those regions and peoples not yet embraced in industrial civilization.

World War II, regarded as an enterprise in the popularization of science, must be counted a success. Well-fed, well-clad troops deploying from lavishly supplied bases, especially American, convinced the denizens of remote and backward places that change is at hand in the condition of man. The "war aim" of Freedom from Want articulated the expectations of hundreds of millions of people drawn suddenly into the twentieth century. For them as for the people of the industrial countries, the war and the bomb that ended it gave a kind of inverted proof of the limitless capacity of science and technology to answer human need. Economic development was established as a primary concern of the United Nations, charged to its numerous technical agencies and to the Economic and Social Council and to the Trusteeship Council that fill boxes on the organization chart coordinate with that of the Security Council.

The United States has contributed generously to the framing of these aims and institutions. Going beyond the traditional renunciation of territorial ambition, the last five U.S. administrations have declared the national interest to be a world at peace and free of want in which the dignity of the human estate is open equally to all men. Throughout this period, U.S. policy abroad has insistently advanced its model of the world community, a community of market economies that achieve growth, prosperity, and natural harmony of interest through freedom of enterprise and trade. Moreover, the U.S. government reckons it has laid out $60 billion (excluding military grants) in economic assistance to other nations over the past twenty years.

About the first $40 billion or so, laid out in the first ten years, there is general satisfaction. These funds went principally to restore the industrial economies of Western Europe. The balance has gone to support the development of preindustrial economies. These outlays have been made with declining enthusiasm and against increasing domestic opposition. Their beneficiaries appear to be in worse trouble than before, and the United States seems to have fewer friends than ever.

In fact, the gap between the industrial and the preindustrial nations has been widening and deepening. Their ways of life are so divergent that food production and consumption provide almost the only indexes by which they can be compared. The disparity before the war is expressed by figures showing that the industrial nations, with 33 percent of the world population, produced nearly 60 percent of the food. In the intervening years, food production has been pushed upward by 50 percent on both sides of the gap. But the population of the industrial countries has grown by only 21 percent, while that of the pre-industrial countries has grown by 52 percent. Since the peoples of the industrial countries were already adequately supplied on the calorie scale, they have converted a large part of the 30 percent increase in their food supply into protein. The peoples of the preindustrial countries have maintained their inadequate calorie intake by increasing the protein deficit in their diets. This is the background for responsible estimates that two-thirds of these peoples, or roughly half of the world population, suffer malnutrition.

The improvement in nutrition scarcely shows up in the economic indexes of the industrial countries. Agriculture constitutes less than 5 percent of the gross national product of the United States even as it yields surpluses for shipment overseas. The production of the entire shopping list of tangible goods now generates, in fact, less than half of the gross national product. As much as 40 to 50 percent of the tangible product is presently devoted to capital investment, to arms, to ventures in space, and to economic assistance without detectable subtraction from the continuous increase in current consumption and with an appreciable portion of the plant and the labor force standing idle. More than half of the labor force is employed as providers of services and nearly half of these in not-for-profit activities undertaken in the public interest. The mechanization of production is carrying the peoples of the other industrial countries in the same direction away from the historic common lot of mankind. It is true, they find new kinds of hazard and discomfort in cities where more than half of them now live, but presumably they will manage these by the exercise of good will and a little of the ingenuity that placed them in this man-made environment.

In the preindustrial countries, agriculture engages 70 percent or more of the population and produces from 35 to 50 percent of the gross national product. Output responds sensitively to demand for bulk commodities from the world market (i.e., the industrial countries) but yields slowly to demand for subsistence from the increasing populations. The gain in food production has been won by bringing poorer land under cultivation and hence with serious damage to resources as well as reduction of the traditionally low yield per man-hour. Scarcity and underemployment on the land have made the cities of these countries the centers of a world-wide population "implosion," crowding uncounted millions of people into squatter communities on their perimeters. The cities may otherwise be regarded as enclaves of the industrial economies to which the countries' exports go; they depend upon imports not only for hard and soft goods but for fuel and even for food. Except where oil and mineral exports bring in large and steady earnings, the foreign exchange balances of all of the preindustrial countries are negative. Out of some 90 preindustrial countries in Asia, Africa, and Latin America, some 30 are under government by military junta.

Subject to qualification, amendment, and exception from case to case, this is the gap between the industrial and preindustrial, the rich and the poor countries that make up the divided world community west of the Urals and south of the Himalayas. Mainland China is excluded from this survey because it is effectively excluded from that community. China is a preindustrial country engaged in industrial revolution. It is proceeding to the task in historically validated style. The traditional deprivation of the peasantry and the urban poor in the distribution of the country's insufficient product has been sharpened by harsh measures of coercion to increase the rate of savings. Investment absorbing 30 percent and more of current production has now secured large objectives in heavy industry, transport, energy, and watershed management. Sober and informed observers declare that the population is no longer exposed to the risk of famine. Within twenty-five years, it is estimated, China will be a first-class industrial power.

China is excluded also because the purpose of this discussion is to explore the question whether the industrial revolutions that impend in Asia, Africa, and Latin America must follow the precedent of the Chinese, the Russian, and the earlier classical industrial revolutions or whether a different route can be opened up. The possibility to be considered here is that economic and technical assistance from the industrial nations might moderate the coercion which otherwise, in conditions of scarcity, appears necessary to the process of capital formation. Industrialization by one route or another offers the only escape from the Malthusian blind alley into which the preindustrial nations have been pushed. Especially in Asia and Latin America, their population growth has outrun the capacity of the resources accessible to exploitation by primitive agricultural technologies. The so-called population explosion ought to be regarded as a first consequence of their acculturation to industrial civilization. It reflects a sharp drop in maternal and infant death rates, and it motivates a readiness for change.

Industrialization is necessary to push the increase of food production ahead of population growth. The vision is not, however, the tractor and combine. The labor-intensive methods that produce such a high yield per acre—5,500 original calories per day in East Asia compared to 2,500 in the United States—can be improved by new implements and power tools but cannot be readily mechanized. What is needed to increase food output is water, to begin with. In the arid and monsoon countries, regulated and dependable water supplies can make two and three crops a year possible where a single crop is now a gamble. Synthetic fertilizers are essential to maintaining productivity in the nitrogen-poor soils of these regions; pesticides are necessary to secure the gain for human consumption. Facilities for storage, processing, and transportation must be provided to deliver the food, especially the more perishable protein foods, to the increasing percentage of the population that is migrating from village to town. All of this implies the installation of the complete apparatus of an industrial system with all of its interdependent working parts.

At the present rate of population growth in Asia and Latin America it will take a 50 percent increase in food production over the next fifteen years just to maintain the present inadequate nutrition. If the rate of population growth is to be reduced, demographers agree, the rate of increase in production must be pushed ahead sufficiently to bring an increase in output per capita. A minimum subjectively appreciable increase in individual well-being is required to motivate fertility control.

The world population is scheduled to double by the end of the century. The next doubling after that will bring it to 12 billion people. Authorities as diverse in their outlook as J. D. Bernal, Colin Clark, and Harrison Brown agree that the world's resources worked by known technologies can sustain a still larger population. But the human species must eventually bring its numbers into stable adjustment with its environment. If there is some optimum, the chance of arriving at it will be improved the sooner production overtakes population. But production cannot even keep up with population growth in Asia and Latin America over the next fifteen years without a substantial start on the task of industrialization.

Development efforts on the scale plainly indicated here were projected with greater confidence around 1950, when the UN technical agencies were first putting world statistics together. With the population estimated to be increasing at 1.2 percent from a base of 2.5 billion, it was reckoned that $500 billion invested over a 35-year period would bring a 50 percent improvement in individual nutrition and an overall 1 percent per annum increase in income per capita. By the end of that period a third of the rural population would have moved into occupations off the land; the population increase would be reduced to 1 percent and the further growth of the new industrial economies would be self-sustaining. The industrial countries were to supply 40 percent of the capital input. At the rate of $6 billion per year this was not an unreasonable expectation.

Today similar calculations would have to start from a bigger population base and reckon with a higher rate of growth. But even if all the magnitudes had to be doubled, the numbers would remain relatively finite. The flow of long-term loans from the industrial to the preindustrial countries is running at the rate of $6 billion a year. A substantial percentage of this is in food and similar relief shipments that do not contribute directly to increase in productive capacity at their destinations. But economies that can sink $120 billion a year in arms could easily manage much larger inputs of capital.

It is now understood, however, that calculations of this kind do not exhaust the complexities of "foreign aid." A large body of literature has accumulated, and development has become a subdiscipline of economics. Among other conclusions, studies have shown that the preindustrial countries cannot absorb capital at a rate significantly higher than the present. Authorities cite the absence of stable political leadership, the shortage of entrepreneurial, managerial, and technical talent, the lack of skilled labor, and the universal illiteracy of the unskilled manpower that is so redundantly available. To discourse of this kind, representatives of the preindustrial countries have answers; for example, Alex Quaison-Sackey, speaking for Ghana before the UN Assembly: "Many bogus sociological arguments are put forward to justify the flow of economic assistance at its present paltry levels. Our country, in fact, would have fruitfully employed capital at many times the rate available. In proceeding on our own, as in the construction of the new port of Tema, we have had to ask great sacrifices of our people and yet reduce our programs below the scale and capacity we know will be required by our economy ten years hence." The fact that Nkrumah is now out of power does not vitiate his ambassador's thesis. Perhaps this is the principal gain of the past twenty years: colonies have become nations with sovereign interests and spokesmen to assert them.

Distinguished spokesmen for both sides are to be found on the ECOSOC advisory committee on the application of science and technology to development. They agree "that the most critical limitation on the capacity of a country to absorb and apply or adapt science and technology to development is its supply of trained manpower." The program they urge cuts between the colonial system that trained up a small elite of civil servants from the native populations and the mass literacy campaigns of revolutionary romance. They call for "increasing the size and balanced composition of the cadres of scientific and technical manpower." This means the creation of schools and specially designed curriculums at the secondary and junior college levels. The graduates will meet the need for technicians, draftsmen, and other middle skills, and from their numbers will come the talent capable of further training. Judging by the impact of reforms in the high school science curriculums of the United States instituted by university professors less than a decade ago, such effort can pay off richly within finite periods of time. The textbooks and teaching materials from these programs are already finding their way into the schools in preindustrial countries.

Of all the tasks in development, this is the one that most explicitly requires external assistance. It is also a contribution that can be made without cost in the sense that the stock of knowledge is not diminished by the sharing of it. The effort, however, takes time and, more poignantly, the lifetimes of rare and devoted individuals. Experience shows that the teacher must go to the students. Despite barriers against immigration everywhere in the world, students who go abroad from poor countries for their training too often fail to return home.

The model for all enterprises in the transfer of knowledge remains the work of the Rockefeller Foundation in upgrading the agriculture of Mexico and other Latin American countries. In Mexico the Foundation established a laboratory and experiment stations that served at once to breed optimum strains of the potato, corn, and wheat plants and to train the full range of personnel, from research agronomists to extension agents, needed to carry these genetic inventions to the farmers. After two decades, with the increase in food production rising to 10 percent per year, Mexico has joined the company of food-exporting economies and the improvement in nutrition is showing up in the population's life tables. The program is self-sustaining under the management of its own alumni, and one of them has served as Secretary of Agriculture in the federal government of Mexico. It serves also to demonstrate that techniques which are the product of generations of accomplishment in the history of science can now yield momentously effective results within a human lifetime.

Service of this kind naturally holds deep attraction for members of the university community in the United States and other industrial countries. American foundations that have sponsored such enterprises have found it necessary, however, to create a kind of civilian foreign service, with professional personnel committed to careers. A leave of absence of a year or two from a university appointment does not give enough time for useful adjustment and performance in the strange environment of a poor country; five years make too large a break in responsibility and advancement at home.

New scales of value and institutional arrangements are required to bring the intellectual resources of the industrial countries into effective connection to the tasks of economic development. One potentially replicable model is provided by the counterpart associations established, with the sponsorship of the Ford Foundation, between universities in the United States and in India. The University of Missouri and the agricultural research and training institute at Bhubaneswar, the consortium of engineering schools headed up by Purdue and the technical institute at Kanpur, and the Harvard Business School and a new school of business administration at Ahmedabad exchange faculty and students and conduct joint research enterprises under arrangements that promote continuity of careers and achievement for individuals on both sides. Whatever the econometrists are able to quantify about the value of these ventures, there is no doubt that they yield the highest return per dollar of all forms of economic assistance.

Something like the same leverage attaches to investment in the next phase in the application of science and technology to development. This is the surveying and primary engineering of resource development. What required more than a century—from Lewis and Clark to the complete 10-foot contour interval, 1:24,000 geodetic survey map—to accomplish for the United States can now be scheduled in months for a preindustrial country with aerial photography and airborne prospecting instruments. Under financial sponsorship of the UN Special Fund, for example, the system analysis for the management of the Mekong watershed is now virtually complete. Sites for the principal dams have been selected and even drilled; and the major distribution works have been laid out to accord with the logic of a soil inventory and land use map. The input in this case of engineering valued at $14 million has set the stage for investment that will eventually total $7 billion. The enterprise is notable also because it has engaged the talent and the national treasuries of eighteen countries, including India and Japan, as well as the United States and the United Kingdom. Elsewhere in the preindustrial countries there are twenty-nine watersheds of comparable square mileage awaiting similar preparation. Meanwhile, in India, the monsoon rains that clear the Western Ghats facing the Arabian Sea run down the Godavari River to flood out villages on the Bay of Bengal.

In the arid lands of North Africa and the Arabian peninsula, engineering enterprises sponsored by the UN Food and Agriculture Organization have opened up the fossil waters of the Pleistocene for use on lands that were abandoned by farmers at the fall of Carthage. The Intercalary aquifers of the Sahara hold potentially greater wealth than the oil reserves discovered there in the past fifteen years. Exploitation of this resource demands the cooperation of the ten nations whose territories it underlies; for the underground flows of the waters, now traced out and mapped, ignore the political boundaries and are likely to prove highly sensitive to the rate of withdrawal at strategic points. It has been suggested that the energy needed for pumping be contributed by the industrial nations that have staked claims to the oil reserves. Water sources of a similar geological vintage have been opened up in Greece and in Lebanon, and the mapping of the fossil waters of the Arabian peninsula is under way. For all of these lands there is in prospect a revival of the agricultural productivity that made the Fertile Crescent the cradle of urban civilization and sustained the four millennia of Classical civilization.

Recently, aerial photography added 40 million acres to the forest reserves of Mexico. In Chile the aerial gravimeter and magnetometer located what has proved to be a 10 billion-ton body of 60 percent iron ore, and in another South American country what promises to be a comparable deposit of copper and related nonferrous ores. These are examples at once of man's ignorance of the resources of large reaches of the earth and of the vital role that the "lead time" inputs of science and technology have to play in the planning of economic development. Considering the objectives and possibilities that knowledge of this kind can open up to national plans, the work ought to be pressed everywhere on a larger scale.

The exploitation of such resources requires large investments in applied research and pilot plant. Because the time course and results cannot be set in advance, it is urgent that work of this kind be carried forward aggressively in order to trigger off the next phase of development at the earliest date. The iron ore reserves of Chile lie far from any coal beds; so do comparable ore bodies in Venezuela and in West Africa. Along with large ore reserves, Brazil and India possess substantial beds of low-grade coal. These ores are moving into world commerce and the blast furnaces of industrial nations. If they are to provide the substance of domestic steel industries in their countries of origin, research will have to be conducted into methods of direct reduction (that is, by hydrogen-oxygen reaction) and, in the case of India and Brazil, the beneficiation of coals laid down before the ice age before last.

In the steel industries of the industrial countries there is no disposition as yet to inquire into these possibilities, even though their own coking-coal reserves are diminishing. The preindustrial countries lack the command of steel technology that must surround such research. Here is an assignment for an international laboratory. The work should not be postponed until the industrial countries run out of coking coal. The preindustrial countries possess many other mineral resources peculiar to the geology of the broken continent of Gondwanaland; the technology for exploiting them is either nonexistent or it is a proprietary asset of some industrial nation.

It has been easier to enlist the sponsorship of industrial countries for research in agricultural technology. The International Rice Research Institute in the Philippines, along with the Indian research station at Cuttack, has successfully crossed the high-yielding *Japonica* with the *Indica* strains native to South Asia and produced many successful varieties of the cross adapted to the diversity of niches in the agriculture of this vast region. In the arid-land countries around the Mediterranean, FAO has sponsored the development of crop-plant strains that thrive in saline soils and brackish irrigation waters. Another enterprise involving collaboration of Israeli and Arab scientists has established a technology for suppressing outbreaks of the desert locust.

There is need for research enterprise on a larger scale and at more fundamental levels to deal with questions of agriculture in the arid-land and monsoon countries that are too big for their national laboratories. Sentiment is growing in favor of the organization of one or more regional institutes. Prediction and potentially control of monsoon storms, management of the fragile laterite soils, an uncataloged bestiary of microorganisms infecting plants and animals—these are some of the tasks on the agenda. Here it should be emphasized that many of the most significant developments in tropical agriculture have come by horizontal interaction among technologists of the preindustrial countries rather than bestowal from industrial countries.

The annual money flows financing the transfer of knowledge in support of economic development—counting bilateral as well as multilateral and UN activities and including such major engineering ventures as the Mekong Valley program—do not exceed $500 million. These outlays take effect, however, through big multipliers. As of January, 1966, the UN Special Fund had committed a cumulative total of $583 million from its own funds to 604 projects (to which the beneficiary countries were committed for an additional $823 million). Out of 73 completed projects, 25 projects representing Special Fund investment of $32 million had generated follow-on investment totaling $1,068 million. That is thirty times the lead-time investment! At half this rate, the projects financed to date by the Special Fund represent a potential $10 billion worth of development. If the rate of investment in development is to overtake the expansion of human need, the lead-time outlays on science and technology ought to increase by at least one order of magnitude. If the preindustrial countries cannot now absorb technical assistance at this rate, the first increases in the rate of investment should go to improve their absorptive capacity. The financial requirement, of course, serves only as an index of the mobilization of people and institutions required. Translated into these terms, the effort can be visualized as scaling out at the size of the U.S. space program. If it could be mounted now, its payoff would come in time to begin to meet the needs of a world population 50 percent larger than today.

The heavier and more visible capital inputs to economic development—such as dams, fertilizer plants, and steel mills—may also be regarded as a mode of information transfer. The principal contribution required of the industrial country is engineering, whether embodied in blueprints or in the shell plates of a blast furnace. Although the multiplier is not as big at this stage, such investment catalyzes large capital outlays, principally in the form of labor, by the recipient economy.

Big programs have become unfashionable in the literature of development, as representing monuments to local leadership rather than effective amplifiers of productivity. But in water management the little works in the valleys cannot function until the big dams are built at the headwaters. A similar argument supports the thesis that industrial development should proceed from the most primary plant that local resources can sustain. With the help of P. C. Mahalanobis, I have calculated that a single unit of foreign exchange invested in the Indian steel industry can generate the value of 75 units of foreign exchange per annum thereafter in food grains. Even if India becomes for a while an exporter of semifinished steel from capacity installed ahead of increase in local demand, as seems likely, the value added will contribute considerably more to relief of the country's critical foreign exchange problem than the export of the country's ores. The same logic justifies the aluminum ingot plant that will soak up the first power generated by the Volta dam in Ghana; the country will be exporting bauxite transformed to aluminum by kilowatts and will earn foreign exchange on both its ores and its falling waters.

The cost of this kind of aid to the industrial economies is misrepresented by the figure that states it in the national budget. In U.S. experience, 80 cents out of each foreign aid dollar has been spent at home. Such expenditure gives the same immediate stimulus to the system as an expenditure on arms or space and with greater prospect of eventual economic return.

The hordes of underemployed people in the countryside and

of the plainly unemployed in the squatter cities have encouraged the idea that development programs should call in labor-intensive technologies. A strong case can be made for the opposite strategy. In the first place, there is a generation or two of labor-intensive work to be done in every preindustrial country on the infrastructure of ports, rails, highways, housing, and buildings that can use up manpower without limit. When it comes to the productive apparatus, however, this ought to incorporate the most advanced developments in science and technology. The model is the petrochemical plant, with two operators at a control panel. Technology at this stage of perfection is highly portable, easily installed, makes least demand on local human resources, and operates at the same efficiency independent of local conditions whether in Galveston or Kuwait. In sum there is no reason why, with adequate capital and technical assistance from outside, the prospective new steel industry in Chile should have to evolve through the beehive coke oven and backyard blast-furnace phase. Ideally, it should install continuous casting at the outset.

The application of science and technology to development may, therefore, offset and reverse the forces that tend to widen and deepen the gap between the rich and the poor. Over the past twenty years the world economy has shown no tendency to approach the natural harmony of its classical model. The preindustrial nations, in their mirror-image relationship to the industrial nations, have found themselves paying higher prices for their imports out of earnings from exports sold at almost constantly declining prices. The international capital market has recorded massive flows horizontally from one industrial country to another and upstream from the preindustrial to industrial countries, but almost none downstream to the preindustrial countries except for investment in extraction of mineral resources. The only counter to these forces has been the flow of economic assistance from government to government in support of development programs and for prevention or relief of famine.

Industrialization, by definition, must revise the relationships now prevailing between the rich and the poor countries. As a developing economy increases the volume and variety of its output, it seeks to substitute goods of its own manufacture for imports and to reduce the relative importance of foreign trade in its expanding total economic activity. This means rigid control over foreign exchange with the aim of diverting export earnings from consumer to capital goods, tariff barriers to protect infant industries, and a variety of other discriminatory practices. In the long run, of course, the developing country will emerge as a bigger and more valued trading partner; by far the largest percentage of the world's commerce flows among the industrial countries. But in the interim the industrial country that provides technical and capital assistance would seem to be acting against its own interest—turning customers into competitors and raising the cost of the commodities it imports from them. Considerations of this kind naturally militate against the flow of aid. In the United States the expansion of the foreign-aid effort is further hampered by ideology. Socialism and related heresies flourish in the planning agencies of the developing countries. There being no private takers for investment opportunities at capital-output ratios of 2.5 to 1 and higher in primary industries as well as in transportation, communication, and utility systems, the planners are pushing for investment in and by the public sector in their countries. What is more, they have to concede that their ambitious public enterprises are unprofitable and can be justified only by the saving of foreign exchange and the external economies they provide for satellite industries. All of this is repugnant to folk economics in rich countries such as our own.

It is now possible to make a meaningful estimate of the capital requirement for development at a rate that will overtake population trends. As the target for the "decade of development" UN agencies set the attainment of a growth rate of 5 percent in the economies of the preindustrial countries. Discounted by population growth this would yield an increase in output per capita of 2 or 3 percent. With the budgets and the foreign exchange transactions of these countries on record and with reasonably solid data available for estimating their national accounts, the objective can be stated in dollars. As of 1960 the gross domestic product of these countries came to $170 billion; their current foreign exchange deficit was $4.9 billion. By 1970, if the decade's objective could be attained, the gross domestic product would grow to $280 billion, and the foreign exchange deficit would be running at $20 billion.

The deficit is the measure of the external capital assistance required to finance the 5 percent rate of growth. It goes without saying that assistance will not be running at that rate by 1970; the gross domestic product will fall well short of $280 billion, and the growth rate will not have reached 5 percent. The need of the developing countries asks historically unprecedented acts of altruism by the industrial countries, especially by the United States. A world free of want can be attained, however, within the lifetime of this generation. That is also an entirely new thing in history. And it has been said to be in the national interest.

VI The Treason
of the Clerks

The Treason
of the Clerks

THE cumulative twenty-five-year outlay by the federal government of $15 billion on university science has financed in America a triumphant era of success and culmination in work on fundamental questions.

The record of the Nobel prizes bestowed on American scientists (62 out of 127 in twenty-five years) does not adequately represent the breadth of the front on which they have contributed to the enlargement of human understanding. To the multiplicity of particles has been added a new multiplicity of natural forces; the biosynthetic mechanism of the living cell has been resolved, clearing the ground for approach to the solemn questions of growth and memory; out of sight on the ocean floor, soundings have disclosed the world-girdling rifted ridge, a topographic feature on the grand scale of the oceans and the continents and the fulcrum from which subcrustal forces move the continents; astronomy, with its reach extended at both ends of the optical spectrum, has come to know a "violent universe," pervaded with processes that proceed at rates and magnitudes never perceived or suspected before.

One cannot begrudge even the university administrators their share in the celebration. Yet, it is necessary to consider the consequences for the autonomy and integrity of the universities of their relationship to the central political power of the nation. And, in view of the present diminution in the flow of federal funds, it must be asked what future has been secured for public support of university science.

Such reflections are forced upon university scientists and administrators by the change in their relations with Washington over the last five years. It is not only that federal support has stopped increasing and has now been overtaken by inflation. It must also be recognized that federal officials are reconsidering the government's relation to the universities. The Pentagon has rudely repaid its most obliging contractors with the report of a study called "Operation Hindsight." It finds that the funds laid out by the Department of Defense in support of basic research in the universities have not contributed to the development of any weapons system or military capability. Considering the effort required in World War II to sell the military on such innovations as radar, this may be sheer myopia. But hindsight now has the force of the Mansfield Amendment, which enjoins the Pentagon to restrict its outlays for research to work addressed to explicit military purposes.

Events have thus nullified the strategy to which the universities and their scientists committed the support of science twenty-five years ago. In the present regressive mood of national politics, the higher learning claims little enough priority put forward at its best. The university goes to Washington today wearing the same tarnish of venality as its detractors there. In the disgrace of disorder at home, with its competence in question along with its mission, it stands in a poor position to make a case for its rapidly expanding needs. Those needs and their priority have no standing because they were never asserted throughout the twenty-five years that set the now ruling precedents for federal support of science and higher education.

Most of the $15 billion came to the universities as incidental spillage from the more than $1,000 billion this country has squandered on its military establishment since the end of World War II. Now that those expenditures—together with the neglect of urgent domestic needs—have begun to stress the essential institutional arrangements of the U.S. economy, they have been called in question. As an equally incidental consequence of this pause for reflection and re-evaluation of national priorities, the universities and their science and engineering departments are in fiscal crisis.

Perhaps there was no reason why university science should not have been financed by the military and paramilitary agencies, much as the late Norbert Wiener and others may have objected. On the other hand, there was no good reason why federal funds for science should have come so primarily from those agencies. Now that the money has stopped, it is clear that the values of science and the mission of the universities have established no claim in their own right on the federal budget.

Belatedly, the university administrators have begun to build the case for that claim. A Carnegie Corporation commission on higher education, chaired by the durable Clark Kerr, has projected the dimensions of the claim. By 1976, according to the commission's preliminary report, the universities will need $13 billion per year of federal funding. The report shows also that some lessons have been learned. It lays emphasis upon education in justifying the $13 billion claim, and it has rather less to say about "service." Showing that important lessons remain to be learned, however, the report "does not . . . recommend any basic change in the present procedure" for the support of science by project contract/grant, "through multiple agencies." Perhaps Project Hindsight and the Mansfield Amendment, published since the preliminary report was issued, will persuade the Carnegie commissioners to re-examine their unspoken assumption that science in the universities can be maintained as a byproduct of services rendered to the military and other mission-oriented agencies.

"The Treason of the Clerks" was presented at the annual meeting, in April, 1965, of the American Philosophical Society. This is the country's senior learned society, founded in 1743 by Benjamin Franklin, who led the clerks of his day on their proper revolutionary mission.

THE intellectual, in Western civilization, has a heritage of treason. No such honorable distinction attaches to the Mandarin, who appears in the long history of China as the servant of sovereigns enthroned by military might, nor to the Brahmin and the scribe communities of India, who hired themselves out to native princes and to Muslim, Mogul, and British conquerors in turn. In Europe, literacy became a revolutionary force even before the invention of movable type. The Western intellectual has ever been a heretic and political dissenter; the subverter, again and again, of the institutions and arrangements of arbitrary power. His most revolutionary enterprise, by far, is science. It is the success of science, manifest in the transformation of the human condition, that has carried the argument for reason against force, for the contract against status in the organization of society. In consequence, one single sovereign holds sway in contemporary political theory; that is, the citizen, robed in all the singularity of his mortal life.

In the half-millennium of social evolution and revolution that has brought us to this time, the university has played a central role. Especially in the last century and a half, it has been the corporate vehicle of the scientific enterprise in its sustained and systematic expansion into each new realm of experience. Today in the Western democracies, the university has the primary function of asserting the autonomy of the intellect, free to seek the truth no matter into what trackless and dangerous territory the search may lead. It is, in this sense, the institutional symbol of the sovereign immunity of the citizen.

In America, the universities have not always fulfilled their constitutional function. Administrators have failed, at times, to hold the gates against intruding vandals. Even before the days of Congressional inquisitions, loyalty oaths, and security clearances, H. L. Mencken could ask without pity: "What would happen to the learned professors [of economics in their most orthodox pronouncements] if they took the other side?" As

Richard Hofstadter has shown, it was the indubitable material accomplishments of the scientists that won immunity in principle for the inquiries of their colleagues in the social sciences and letters. And it is action by professors, organized in their extramural guilds, that secures the ultimate sanctions protecting academic freedom in this country today. Nonetheless, the university remains, in Thorstein Veblen's words, "ideally and in popular apprehension . . . a corporation for the cultivation of the community's highest ideals and aspirations."

There is more than intramural interest, therefore, in the observation that the American university has found its way, over the past twenty-five years, into radically new relationships with the rest of the community, especially with the agencies of the federal government. This is true especially of the one hundred largest universities and more especially of the twenty most prestigious, the ones that set the mold of the American university in general. On the record of the past, one should not be surprised perhaps to find that the internal organization of these institutions has been weakened by the pressures to which they have been exposed. It appears, however, that internal disorganization has correspondingly compromised their autonomy. Still more alarming is the testimony of famous scholars and university presidents who deprecate, justify, or celebrate the disorganization of their institutions and the consequent subornation of the university from its essential role in our society.

Our universities have not recovered, perhaps, from the almost total suspension of university life in World War II. The story of their work in those years presents a measure of the great store of capital of the spirit with which they entered national service. The universities transformed themselves into vast weapons-development laboratories. Theoretical physicists became engineers, and engineers forced solutions at the frontiers of knowledge. The Massachusetts Institute of Technology and Harvard University undertook to create the strategy and tactics as well as the instruments of radar and counterradar; from Johns Hopkins came the proximity fuse that brought the conventional high-explosive artillery shell to its peak of lethality, and Columbia, Chicago, and California joined in the successful engineering and manufacture of the most fateful weapon of all. The universities and the scientists then dealt with the military not as contractors, but as privateers bringing along their own novel weapons; Don K. Price has observed, ". . . it became apparent that what scientists discovered by unrestricted research might be of greater military importance than the things the military officers thought they wanted—in short, that the means might determine the ends."

At the end of the war radical innovations in technology stood ready to carry new advances in the scientific enterprise from which they had come. To seize the opportunities opened up by prospective giant accelerators, reactors, and radio telescopes, the physical sciences had to scale their budgets upward by several orders of magnitude. The investment required to build such instruments dwarfed the outlays for the 200-inch Hale telescope and the 184-inch cyclotron at Berkeley, the last big instruments built on private capital. It was apparent that the public, primarily through the federal government, would have to finance the work of science.

Even before the war ended, scientists and public officials opened discussion on the design of a national foundation for science. The legislation which at last created the National Science Foundation in 1950 was caught in the cross fire of Congressional debate about the atom. But the authors of the foundation also generated a controversy of their own. That controversy was thickened with bad omen for the future relations of the universities and American society.

Leading figures in the scientific war effort, joined by eminent university officials, proposed to place the foundation in the control of a board of part-time trustees. They argued for this private trusteeship on grounds of disdain for "politics" and fear of federal "control." From the same general quarter came a similar proposal for the design of the agency that was to administer hospital construction funds under the Hill-Burton bill. Against that arrangement Senator James E. Murray of Montana wrote: "The [proposed] Council . . . is accountable to nobody, responsible to nobody, controllable by nobody. . . . It is the genius of our American democratic government that responsibility and accountability go hand in hand with power and authority."

President Truman called the turn with his veto in 1946: "The proposed National Science Foundation would be divorced from control by the people to an extent that implies a distinct lack of faith in democratic processes."

The aborted National Science Foundation was not greatly missed. The military services remained firmly convinced of the utility of science and eager to prolong their wartime intimacy with the universities. Under enlightened management, the Office of Naval Research set imaginative precedents for the cutting of strings on funds advanced to support basic research in fields remote from strictly nautical concerns. There was, besides, the United States Public Health Service, which found Congress pressing ever larger research appropriations upon it. In twenty years from 1945 to 1965 the annual federal outlay for what is now called "R and D" has mounted from $500 million to $15,000 million. The more than $1,500 million now flowing into the universities from federal agencies constitutes 15 percent of all expenditures in United States institutions of higher learning and fully 75 percent of the expenditures for research in those institutions. Military and paramilitary agencies—the Department of Defense, the Atomic Energy Commission, and the National Aeronautics and Space Administration—supply 65 percent of these funds. Taken together with the outlays of the Public Health Service, the outlays of mission-oriented agencies make up 90 percent of all federal expenditures for research in the universities. The National Science Foundation, the single agency dedicated to the support of fundamental investigations motivated by the aims of science as traditionally and conventionally understood, swings about 7 percent of the total.

Gigantism at the sources of the federal R and D money flows is matched by aggrandizement at their destination. The top ten universities on the list of recipients soak up nearly 40 percent of the funds; the top twenty, nearly 60 percent; and one hundred universities account for all but 10 percent of the funds. Beginning with Harvard, the list of twenty includes the richest and grandest privately endowed universities; it also includes some of

254 The Acceleration of History

the most distinguished state universities. Federal funds supply
from 32 percent (Harvard) to 85 percent of their *total* budgets,
and from 75 to 100 percent of their research budgets. Whatever
their original identity, they must now be reckoned as constituting a new class of institutions, rightly labeled by Clark Kerr as
"federal grant universities."

A full third of the federal money flow sluices into the fourteen major and the many lesser research institutes and laboratories which the big universities operate for the federal government. Over the past twenty years these "university-associated"
institutes have carried through such assignments as the fashioning of the thermonuclear weapons, the perfection of instruments for detection of nuclear explosions, the design of the
succession of continental bomber and missile defense systems,
the evaluation of weapons systems and strategies predicated
upon them, studies of modes of civil defense, and economic and
general intelligence studies of putative or prospective enemy
nations. Even with the budgets of the institutes set aside, the
flow of federal funds to the twenty federal grant universities
makes up the commanding percentage of their total expenditures as universities "proper."

Federal money comes to the universities proper almost entirely through the instrument of contracts or grants for research
projects of stated short-term duration. Few administrators or
scientists are able, or care, to recall the days when money of this
kind funded a junior and supplementary portion of the budget
of a university department. It is true that statutory and other
regulations set a short term on the duration of both grants and
contracts. In practice, however, this is of small concern. The
money keeps on coming, whether by renewal of the original
instrument or the writing of a new one; no one distinguishes
nowadays between "soft" money and "hard." Nor is there
much point in flogging the distinction between a contract and a
grant. The committee of the National Academy of Sciences
that most recently reviewed the whole question of federal
support for basic research in institutions of higher learning
could dismiss the issue in the comfortable embrace of a pair of
parentheses, speaking of "support by research project grants and
by fixed-price research contracts (not too unlike grants)."

Opinion is unanimous that "research project grants and contracts should remain the backbone of federal policy in support of basic research in science in universities." This is what the committee of the National Academy of Sciences found; the same conclusion has been propounded by every committee ever convoked on the subject and by leading scientists and university presidents in distinguished lectureships, in presidential addresses to educational and scientific societies, and in testimony to Congressional committees. In a livelier discourse than most, McGeorge Bundy has compared federal money favorably to "the over-administered grant . . . by the large private foundations" and to alumni gifts for "the subsidy of athletes or the construction of pretentious and egocentric memorial buildings." The federal dollar, he declared, "has been . . . as good as any other—always excepting the wholly unrestricted gift."

No one any longer argues the virtue of poverty on private endowment or doubts the need for public support. Nor is there any question that federal money has financed a period of towering achievement in American science.

Yet certain hard questions remain. The questions concern the terms of public support. These questions recur, again and again, in the briefs that speak for the unanimity of opinion on the system, but they go unanswered and, some of them, unasked.

To begin with an unasked question: Why should the pursuit of knowledge as an end in itself be financed to the extent of more than 90 percent by mission-oriented agencies? Either the universities should not be doing business with them on this scale, or these agencies are supporting science in an utterly disinterested way and the money would come more appropriately from the National Science Foundation. In fiscal 1964, the National Science Foundation laid out $115 million in 3,105 project grants against 6,020 applications totaling $448 million. In granting half its applicants one-quarter of the funds they requested, the Foundation improved the rate of cash flow by holding the average duration of its grants down to two years. Meanwhile, the Department of Defense disbursed about $400 million to the university-associated institutes and the universities proper, in much larger unit sums and on contracts of

longer duration. Now, the Department of Defense may reasonably argue its need to be able to stir up interest in fields of basic
research neglected by universities. But why should it supply
nearly one-quarter of all federal funds going to university
science? Why, for example, should the Department of Defense
in the academic year 1959–1960 have supplied half of all the
federal funds disbursed to the current support of research in the
Faculty of Arts and Sciences at Harvard? There is no cause,
perhaps, to worry about Harvard, but other universities are not
so rich in tradition or financially independent. Some comfort
can be had from the succession of figures that shows the Defense share of the federal science budget going down. What
Defense yielded on percentage points as the budget doubled in
the last four years went to finance the ventures of the National
Aeronautics and Space Administration. Without any particular
help from the universities, meanwhile, the National Science
Foundation has held on to its 7 percent share.

The new "condition of mutual dependence between the
federal government and institutions of higher learning" is vexed
by many secondary issues. Each, in its way, goes to the heart of
the matter. There is the problem of overhead or indirect costs.
"Ideally," it is declared, "the grant [and presumably also the
contract] makes only a partial contribution toward a purpose to
which the grantee institution is already fully committed."
Somewhat less ideally, it is argued that since the work is undertaken in the government's interest "the no-gain-no-loss principle
for research contracts indicates that the government should
defray these [indirect] costs which . . . [have become] a real
drain on the institution."

The same moral conflict between the motives of the benefactor and the beneficiary underlies the "current trend toward
introducing into grant and contract negotiations . . . administrative restrictions that are inimical to effective basic research"
and the "recent trend toward unnecessary restrictions on scientific freedom and increases in the bookkeeping chores of scientists in both grants and contracts." The conflict becomes explicit in the prayer that "only a deviation from the broad
objectives of a project proposal . . . should be considered as

constituting a change in the purpose of the grant, thus calling for special approval from the federal agency." To the credit of the agencies it is said that "they bend over backward in favor of academic freedom." The trouble is that applications from the investigators "sometimes exaggerate the potential practical achievements of the basic research they propose." Never settled is the issue whether the money is advanced and accepted in the purchase or the support of research.

No self-evident set of principles commends the "project contract/grant" as the instrument for the public support of science in America. What saves discussion about principle is a single reassuring fact: all but about 5 percent of the federal money for science in the universities proper is disbursed on the recommendation of the university scientists themselves. On innumerable panels—representing traditional disciplines and ad hoc undertakings—they take turns sitting on research proposals from workers in their field of common interest, not excluding themselves. The award, therefore, is made on the scientific merit of the proposal, and the applicant has the satisfaction of knowing that he has been judged by his peers. Nor does he begrudge time for service as a panel member, for the plurality of sources assures continuity of support to his field.

The scientist, secure in his knowledge of his own integrity, thus finds the project contract/grant system a congenial one to work with. The university administrator shares his satisfaction, providing overhead costs are adequately funded. Presumably, now, the public should ratify the arrangement and make the satisfaction general. It is only recently, however, that the public, in the person of its representatives in Congress, has become aware of the scale and ramification of the federal patronage of science in the universities. Even the $1,500 million outlay of 1965 constitutes a junior percentage of federal R and D expenditure, and that figure is incidental to the huge appropriations for the arms and space races that usually clear the House and Senate with a minimum of scrutiny. The first hearings on the project contract/grant system were held only a year or so ago, and they recorded little more than the assurance from the beneficiaries that this is the best of all possible systems, pro-

vided Congress keeps the money flowing. Not knowing what
the public will think when it has all the facts, public officials in
the granting agencies are constrained to raise secondary questions, just to clear their own place in the record. At the bottom
of each of these questions remains the big question whether
universities should do business with the public this way.

It may be embarrassing, for example, to have to explain some
day why most of the money disbursed by the scientific panels
has gone to the universities whence the panelists come. The
first thought suggested by this statement can be discounted at
once: the 20 institutions in question are held, by every standard, to be the foremost in the land. On the other hand, the
panel disbursements have made the rich richer and the poor
poorer; the share of those twenty institutions in federal outlays
climbed to its present 60 percent from only 32 percent in 1948.
The rest of the 1,700 colleges and universities have sustained an
absolute as well as relative decline in their income from federal
sources. No one is pushing very hard the proposal that this
imbalance be redressed by "developmental grants" to neglected
but worthy institutions, calculated to build them into new
"centers of excellence" for the receipt of project grants from the
"permanent interrelated system." The liberal arts colleges are
advised, with some condescension, to stay out of the competition for project contract/grants which "can easily subvert them
from their primary obligation. . . ."

The panels have served as positive feedback amplifiers in
another equally significant sense. There was a cheerful irony in
Warren Weaver's description of the "Special Committee for
X"—"These are men intensely interested in X, often with lifelong dedication to X, and sometimes with a recognizably fanatic
concentration of interest on X. Quite clearly, they are just the
authorities to ask if you want to know whether X is a good
idea."

Alvin Weinberg puts the matter bluntly: "The panel system
is weak insofar as judge, jury, plaintiff and defendant are usually
one and the same." Panels and their budgets tend to be self-
perpetuating; the unit outlays tend to become larger and the
number of contracts and grants smaller. Paul Weiss has said

that biological research "has grown softer by loss of self-restraint." Yet the number of sound proposals coming forward in the life sciences, especially from younger workers, exceeds the funds available. From other quarters come complaints that fields not in the purview of the mission-oriented agencies suffer financial neglect. As in the national economy, "the boom is at the top." Federal funds are generously available to the clientele of the mission-oriented agencies, especially the military and the paramilitary agencies; there is dearth elsewhere in the community, where science looks for its support to the National Science Foundation.

The project contract/grant must stand scrutiny not only as an instrument for the support of science, but also as a device for supplying federal aid to higher education. Its function in this regard cannot be minimized by pious disclaimers to the effect that it has served merely "to encourage, in fields colored by national interest, research which our faculty members wished to undertake." That research now engrosses 50 to 100 percent of the budget and energy of the faculties of science. If it belongs on the campus at all, then the university has long since assumed the obligation to see to its support. For that support the big universities are now almost wholly dependent upon federal aid. Not more than one of them can claim: "If federal funds were to be cut off tomorrow, Harvard would be able to honor its commitments to all its permanent faculty members." In addition the universities have been so successful in pressing the contracting and granting agencies for reimbursement of their indirect costs that federal payments on this account constitute one of their major sources of general income. Project contract/grant money has flowed in such huge volume for such a long time that it must now be acknowledged for what it truly is: a subterfuge for federal aid.

The rule that has prevailed in the generation and distribution of this public subsidy is expediency. Because federal funds were most readily available for science, it is science that now flourishes on the great campuses. The university science budget has grown out of all proportion to that of other realms of learning and most disproportionately in those universities that have ·

captured the biggest contracts and grants. Of course, modern science requires larger outlays per investigator than, say, medieval French literature. The sciences appear to have acquired some of their new bulk, however, at the expense of the humanities. This view is pressed by advocates of a National Humanities Foundation, although with more strong feeling than hard evidence. The imbalance in federal aid has had a more serious consequence, however, in the opinion of Harold Orlans, who has made the most extensive on-site study of the situation. "Government research programs," says Orlans, "have cut university liberal arts faculties in two, in terms of income, teaching-load and, most important, status and *esprit de corps*."

Income differentials have been tolerated in the university system before: it is fortunate that most medical schools are located so far from their home campuses. But now this divisive issue reaches into the heart of the university. In the generally more rapid progress of a career in science, a man gains his doctorate two years ahead of his contemporaries in the social sciences and four years ahead of those in the humanities, and he widens the gap still further in his passage through the ranks to professor. Along the way, as compared to those of equal rank in other departments, he enjoys an income differential from such extramural sources as the now "customary" summer salary on federal contracts and grants, from consultantships, and from equities in business ventures that bloom from research. His career may be crowned early by a "distinguished" professorship —a kind of life peerage in which he will have no successor— carrying a higher primary stipend and created to outbid the blandishments of another campus. This situation has not only attracted the vulgar attention of Congressmen, who talk darkly of "conflict of interest"; there has also been discussion in the American Council on Education of the desirability of a "code of ethics" for professors!

In detaching the sciences from their no more than dimly sensed identity with arts and letters, the pressures of federal funds have also loosened the ties that bind the scientific faculties to their universities. The enterprising investigator whose program and ambition command federal support has made the real decision that University Y would win distinction as a center for the investigation of X. Because he must be chronically engaged in seeking renewal of his project contracts and grants, he comes to think of the granting agencies as his alma mater, his "true source of nourishment," and to identify himself more closely with his colleagues and competitors around the country than with his fellow faculty members.

The disengagement of the scientist's loyalties finds harsh economic expression in his emergence as a third party to the interminable haggling about indirect costs. Knowing that the national panel in his discipline has only a finite number of dollars to grant, he regards payments to the overhead of his university as deductions from his grant and he opposes them. After the fashion of Madison Avenue account executives, he is in a position to take his staff and his grants elsewhere if the university does not meet his terms. Administrators thus come under pressure to make concessions damaging to the morale of their faculties in order to attract investigators bearing "large-scale research projects of high visibility." As one university president, celebrating a substantial benefaction from the Atomic Energy Commission, declared to his regents: "The key to success is the faculty member who can and does attract grants, contracts and other financial assistance."

It has been suggested that competition for talent of this kind might be tempered by a code of ethics for university administrators. Deploring "the flight from teaching," the trustees of the Carnegie Endowment for the Advancement of Teaching declared: "Leading universities might agree among themselves to exercise restraint in offering reduced teaching loads as an inducement to move." A "teaching load," of course, is something borne by university faculties; the term does not come from the liberal arts college. For more than a generation, undergraduates in the university colleges have complained about the poverty of their exposure to the eminences of the faculty. In the universities most richly supplied with project contract/grant money, however, the natural scientists, and the social scientists as well, have now reduced their undergraduate teaching hours well below those of their colleagues in the humanities and to a still smaller fraction of the hours taught by professors of both cultures in the liberal arts colleges. The contract document, in fact, usually stipulates that the principal investigator will devote a stated percentage of his time to the project and that federal funds will not be used to subsidize his teaching time. All of the criteria—from the number of National Science Foundation Fellowships bestowed upon graduating seniors to the number of Ph.D.'s conferred per 1,000 alumni—indicate that the liberal arts colleges, rather than the university colleges, are recruiting the next generation of scientists. Moreover, the list of the first ten federal grant universities does not correspond to the list of the first ten producers of doctorates in the sciences. The neglect of graduate teaching, presumably in favor of research, must inevitably reduce the quality of the research; for it is well known that the interplay of master and student in advanced teaching takes the research into its most unexpected and fruitful turns.

The principal casualty of the federal grant system, however, is the undergraduate. In his Godkin lectures at Harvard in 1963, Clark Kerr said:

Recent changes in the American university have done them little good—lower teaching loads for the faculty, larger classes, the use of substitute teachers for the regular faculty, the choice of faculty members based on research accomplishment rather than instructional capacity, the fragmentation of knowledge into endless subdivisions.

To all of which Kerr added—with some prescience—"There is an incipient revolt of undergraduate students against the faculty. . . ." When the revolt came, however, it was against the administration.

To the traditional functions of the university—teaching and research—the federal grant universities have added a third: "to serve the community and the nation directly through its faculty and through the use of its material and administrative resources." The teacher and researcher has been joined by a new personage, "the consultant and administrator." Under his leadership, embracing the third function, "the university has become a prime instrument of national purpose." With a backward glance upon "the quieter scholarly domain," hope is expressed that "erosion of traditional academic values" may be kept from "stifling the fruits of basic research." But the "arena of action" pulls hard against the "scholarly vacuum"; the "pursuit of knowledge for its own sake" must acknowledge the complementary attractions of "knowledge in the service of mankind." So the university must be "drawn increasingly into more direct participation in the affairs of society" and "merge its activities with industry as never before."

Our greatest universities have now been "in the mainstream of events" for more than twenty years. It is not, therefore, really so "incongruous" that "they should be charged with so large a measure of responsibility for defense research." To launch new centers for research in this field in competition with them has proved to be no easy undertaking, whether for profit or not for profit. When the start-up of new projects in the arms race yielded to the mobilization of high technology for the space race, the great universities were there to command contracts and grants from this enterprise, regardless of its desirability or the priority of its claim upon the nation's scientific resources, rationally and humanely allocated. In a hopeful vision of the future, the universities "will never again be able to isolate themselves wholly from enterprises that bridge the interests of the scientist and the scholar with the world of action": "the transformation of our cities," "the crisis in transportation," "the shift to automation in industry," "the highly controversial use of pesticides," and, even, "the progress of education itself."

Meanwhile, the social sciences have also begun to win recognition. The new science of "human engineering" at the "man-machine interface" has brought psychology into the circle of disciplines favored by the project contract/grant. Regional research institutes, organized at the primary initiative of the Department of State and the great private foundations to illuminate hitherto dark regions on the world map, have brought sociology and anthropology into the ambience of the Department of Defense. Lately, even the humanists, who had previously been confined to such servile chores as consulting on official histories of the last war, have found more active assignments in "area and language training for military personnel and studies of certain strategic peoples." With funds abounding for projects in every field of learning, the university campus has come to harbor a new kind of condottieri, mercenaries of science and scholarship hooded with doctorates and ready for hire on studies done to contract specification. Studies of this kind have been solemnly entered in the records of Congressional hearings, released as reports to federal executive agencies, and published by university presses.

Under the circumstances, one ought not be surprised to learn that an organization called the RAND Corporation has been able to "stimulate the academic environment." One should be prepared equally to understand why, in so many instances, "the scientific community has found the surroundings it needs for outstanding work within the walls of both governmental and industrial laboratories as well as in the universities."

As instruments of national purpose, the federal grant universities have avoided the hazards of politics and eluded governmental control. To a greater or lesser degree, however, each has compromised its integrity as a university.

The expansion of project contract/grant research has downgraded the teaching function in all of these universities—with but a few notable exceptions—especially in the sciences and including even graduate education. The faculties of arts and sciences are divided by the meanest issues of compensation and status, and the scientific faculties owe allegiance to powers outside the university community. The character of university science in America is clouded by the question of whether it is conducted in the interests of the contracting and granting agencies or for the universal, enduring end of increase in knowledge. In any case, because these universities cannot guarantee the long-term support, they cannot defend independence of their scientific faculties.

In all of these respects, the federal grant universities have weakened their autonomy and so their capacity to fulfill their function as the corporate agents of free inquiry. They have explicitly surrendered this function to the extent that they have undertaken to serve as the executors of federal policies and programs, for to that extent they have compromised their standing as centers for the independent criticism and surveillance of those policies and programs.

The citizen has never had greater need of the protection afforded by the university than in the present epoch of centralization of power in the national state. But our great universities are engaged in abrogating their commitment to his dignity and liberty. After a half a millennium, the clerk now turns in treason upon the sovereign he installed in power.

The project contract/grant system has not settled the question of public support for higher education. In fact, the chorus of endorsement from its beneficiaries comes at the time when the dangerous defects of the system are becoming apparent. In 1963 Jerome Wiesner, from his central vantage point as science adviser to the President, concluded: "The time has come to find a means by which the universities can accept more of the responsibility for the allocation and use of the funds." He added: "I suspect that neither the funding agencies, the universities, nor the individual researchers will welcome this suggestion. . . ."

The need for public funds embraces larger realms than science and requires larger flows. Increased enrollments, increased costs of education, the need to upgrade the quality of instruction even while reducing the economic, racial, geographical, and other barriers that separate able youth from educational opportunity, cumulatively demand steep multiplication in the total national outlays to higher education. The university colleges, in particular, must readdress themselves to their obligations. They provide, in theory, the ideal environment for the self-discovery of the future scholar and scientist; the most promising young people in the land are now attracted to them. In this regard, the colleges of the great private universities have special needs; it has been estimated that not more than 24,000 members of this year's high school class who qualify for college admission with a Scholastic Aptitude Test score of 550 or better can meet the mounting cost of education there without financial assistance.

The universities must at last seek openly the public support they need. By attempting to arrange inside and private solutions for their financial problems, they have expressed an unbecoming contempt for the electorate. They cannot write off the democratic process as one that will bring, "after a period of general logrolling, the enactment of a statutory formula to give aid to all institutions equally, or some formula based on population or geography, regardless of scientific or academic standards." Better public understanding of the high public function of the university will win due respect for academic and scientific standards in the allocation of public support. By promoting such understanding, the universities will fill a significant gap in the higher education of a society in which 10 million college graduates are now at large.

Federal Funds and Science Education

EVEN at the level of $1.5 billion per year, federal funding for science in the universities did not carry enough political romance to command the attention of its own committee in the Senate or House. Science turns up explicitly in committee titles only as the junior preoccupation of the House Committee on Science and Astronautics, much as it used to find its way into the high schools in the classroom newspaper called Science and Aeronautics. As chairman of the Subcommittee on Science, Research and Development, Emilio Q. Daddario provided the one Congressional forum in which questions of public policy in science were consistently and constructively pursued. His retirement from the House to seek, unsuccessfully, the governorship of Connecticut breaks an important link between the scientific community and the national legislature. This essay was contributed to the committee's proceedings in January, 1966.

Over the past two decades the federal government has become the nation's principal benefactor of science. On appropriations each year bigger than the last, science has flourished in the universities of America. Now, with the Higher Education Act of 1965 in force, the government is about to take up the same role with respect to the university system as a whole. As a first consequence, the beneficiaries of the open-handed support of science must anticipate significant revision in the terms on which public funds flow to them.

The clearest forecast of the shape of policy to come is offered by the statement "On Strengthening the Academic Capability for Science Throughout the Nation" issued in September, 1965, by the President to the heads of departments and agencies. The President observes that federal outlays for science in the colleges and universities—". . . where," he says, "education and research become inseparable"—presently add up to $1.5 billion and account for two-thirds of the total research expenditures of these institutions. "Plainly the Federal expenditures have a major effect on the development of our higher educational system." The statement concludes: "Our policies and attitudes in regard to science cannot satisfactorily be related solely to achievement of goals and ends we set for our research." Department and agency heads are accordingly instructed to manage the disbursement of research funds in such a way as "to insure that our programs for Federal support of research in colleges and universities contribute more to the long-run strengthening of the universities and colleges so that these institutions can best serve the nation in the years ahead . . . [and] to find excellence and build it up wherever it is found so that creative centers of excellence may grow in every part of the nation."

This statement of policy responds tacitly to disquiet that has been growing in the Congress over the past two years, as this committee and others have put together the overall picture of the disbursements and the practices of the numerous granting agencies that make up the total federal purse for science in the universities. Evidence has come in to support concern that federal funds have unwisely favored some fields of science as against others, some institutions as against others, and some regions of the country as against others; that emphasis upon research has diluted the quality of teaching especially in the colleges, and that perhaps the build-up of science has disadvantaged other realms of scholarship. Expression of such misgivings has come principally from members of Congress and others outside the academic community. On all these headings, scientists and university administrators have been at pains to offer reassurance in their testimony before this committee and in their contributions to the growing literature on public policy in science. They will apparently accept with reluctance any revision in federal granting practices along the lines projected in the President's policy statement.

It may seem curious that educators should want to discourage governmental concern for the well-being of education. The underlying motivation for this attitude, however, deserves respectful consideration.

Education engages the crux of the paradox of citizenship in a self-governing democracy. The citizen is at once the governed, and so subject to the power of his government, and the sovereign governor of that power. In the eighteenth-century vision of our society, it is the task of education to perfect him for his dual role. Anything that touches on the relationship between the government and the educational system, therefore, touches the central nerve of democracy. At the apex of the system, in the university, the sensitivity is at peak. The government has all kinds of uses for the university. If the national defense requires thermonuclear technology or if the general welfare calls for increased numbers of engineers, where else shall the government turn? But the university, rightly conceived, can respond to demand for service to the ulterior motives of the government only to the extent that such service comports with its prior obligation to the citizen. For the university is the seat of the citizen's sovereignty; it is the institutional embodiment of the immunity that hedges his liberty. In the realm of scholarship and research, no external authority can make policy for the university. It is by his absolute freedom to inquire that the citizen as governor makes the policy of his government. His inquiries, especially in the realm of science, have changed the course of history. The success of science popularized by the transformation of the material condition of man during the past two centuries has brought corresponding transformation of the relation of man to man. The self-government of the republic is no longer reserved to the patrician minority; the sovereignty of citizenship is open to all members of society.

These are the considerations that underlie the citizenry's original directive to the federal government on the subject of education: Hands off! George Washington's proposal to establish a national university was buried by the Congress. The funding of public education was reserved to the smallest and weakest units of government. Today, in suburban America, the meeting of the school board remains as the last vestige of the direct democracy of the town meeting. Higher education was at the outset reserved to private enterprise. Down to this day, the autonomy of the state universities has been sheltered by the prestige of the great private universities of America.

It was the towering material and financial requirements of modern science that forced the entry of federal funds and federal policy into the realm of higher education. Nothing less than the national treasury can finance the giant instruments and undertakings of big science. The same banker is the only one that can respond with sufficient speed and flexibility to the burgeoning enterprises of little science.

No one doubts that the citizen as taxpayer has got his money's worth in return for his generous patronage of science. But, great as his satisfaction ought to be, this does not ratify all the policies, terms, and practices of the federal granting system.

That system must stand up to the question whether it is the best that can be designed to secure the maximum return on the public's investment. Even those who have reacted so anxiously to indications that the system is about to undergo revision concede it has its imperfections. These they discount, however, on the pragmatic ground that the system minimizes the perils of political interference and governmental control. No single agency holds an exclusive franchise on the granting of funds, and so none has the monopoly of power that would go with it. All but about 5 percent of the funds are disbursed by panels of scientists recruited from the field to which the funds are appropriated; awards are made, therefore, on the basis of scientific merit and merit alone.

Now, of course, independence of the scientist and the autonomy of the university are the most important considerations of all. The system must withstand scrutiny on this score as well. As in other human activities, questions of principle here seem to lead most directly to the heart of the matter. This, incidentally, is perfectly good pragmatism; for principle may be understood to be the distillate of experience.

The first question of principle is raised by the finite and brief duration of the project contract or grant that delivers the funds. Society through its universities undertakes to support the scholar for the duration of his active life. In fields other than the sciences the university's own, or "hard," money substantially covers this undertaking. Project grants have been known in university science from the days when the great private foundations furnished the principal external support. Such funds constituted, however, the junior percentage of the (inadequate) total funding. Federal project funds today hold a two-thirds interest in total university expenditures for research and a still more commanding interest—exceeding 80 percent—in the research activities of the 10, 25, or 100 universities that receive most of the funds.

With such a high percentage of the total activity hanging upon project-by-project support, the independence of the university scientist would seem to be heavily compromised. But so long as appropriations grow bigger every year the question must seem so old-fashioned as to be crotchety. University scientists stopped complaining equally long ago about the nuisance of promoting funds and sitting in judgment on the disbursement of funds. As a matter of principle, however, they should take note that they do not have a claim on the national treasury. Until they have such a claim, asserted and established on the true merit of their function in society, their work starts, stops, or continues at the will and convenience of others.

The hazards implicit in discontinuous project support are amplified by the fact that more than 90 percent of the funds are supplied by mission-oriented agencies. More than half of the total comes from the military and paramilitary agencies. Strictly speaking, the charters of these agencies permit them to purchase research, not support it, and only such research as may advance their mission. After precedents established by the Office of Naval Research in the first postwar years, however, Congress has encouraged the agencies to construe their missions with sufficient breadth to invest in basic research. The universities accordingly feel free to regard the research contract of the mission-oriented agency as indistinguishable from a grant by the National Science Foundation, the one agency chartered to support basic research.

The citizen, concerned for the autonomy of the university, must nonetheless ask why the defense agencies should provide more than half of the support of basic research. Survival is a compelling preoccupation. But the simpler answer is that these agencies have funds in abundance, while the National Science Foundation does not. As Alvin Weinberg has observed, "The tendency of mission-oriented agencies to do basic research that is relevant, though only remotely, to their missions has in a way defeated the original design for the National Science Foundation. For now we have eight or nine 'National Science Foundations'—the National Institutes of Health, the Atomic Energy Commission, the Armed Services and others." Weinberg, "along with many others," argues "against putting support of all science in one compartment of the budget." To the National Science Foundation they assign the "balance wheel" function of picking up the slack when one or another mission-oriented agency falters.

The preference for a plurality of supporting agencies explains in large part why the scientific entrepreneurs of the universities were so quick to embrace the space program, even though the priority accorded to this adventure is questioned by many of their colleagues. For a while, the increase in space outlays offset the relative decline in the defense account. It appears, however, that the leveling out of space research now threatens a financial crisis in the physical sciences. If the National Science Foundation is to take up this slack, it must become something more than a balance wheel. University scientists will have to embrace at last the long-overdue task of persuading their fellow citizens of the wisdom of supporting science for its own sake.

The expenditures of the Department of Defense, the National Aeronautics and Space Administration, and the Atomic Energy Commission greatly exacerbate the concentration, noted by this committee, of federal funds in a relatively few institutions. While some one hundred universities absorb 90 percent of the total federal outlay, the ten largest on the list of Department of Defense contractors receive 70 percent of defense outlays. The ten largest on the National Aeronautics and Space Administration list receive nearly 60 percent of that agency's outlays, and the ten largest on the Atomic Energy Commission list, nearly 50 percent. In contrast, the top beneficiaries of the National Science Foundation receive less than 40 percent of that agency's disbursements.

Mission-oriented outlays similarly accentuate the disparities in the geographic distribution of funds. As this committee has noted, the North Central and South Central states, which turn out respectively 31 percent and 14 percent of the nation's Ph.D.'s, receive less than their share of the National Science Foundation outlays, 25 percent and 8 percent respectively. They share even more unfavorably in the total federal flow: only 18 percent and 5 percent respectively. The mission-oriented funds follow the distribution of defense outlays rather than the distribution of educational facilities into the Northeast and Pacific regions, with 30 percent of the funds going to the Pacific states. Nor is the picture much changed if one eliminates, in accord with Lee DuBridge's urging, payment to the university-associated research centers. Mission-oriented funds tend to flow in highest concentration to an agency's contractor in the management of one of these centers. The twenty-five top private (fifteen) and state (ten) universities received 60 percent of all the government expenditures on university science, and a still larger percentage of the mission-oriented outlays. In sum, the concentration of federal funds has indeed tended to make the rich richer.

Whether these funds have made them truly richer remains to be seen. DuBridge is the author of the estimate that less than one-fiftieth of the nearly $17 billion federal outlay for "R and D" is spent on basic science. Something less than $350 million of the $1.5 billion in federal funds supplied to universities must, therefore, go to support their primary mission. This leaves more than $1 billion to be accounted for by other undertakings in the universities, including the relatively lumpy appropriations to big science, not reckoned as "basic" in DuBridge's strict accounting. Given the concentration of the mission-oriented funds, most of these undertakings find their setting in the twenty-five foremost universities. On some campuses, the rightful denizens are beginning to suspect that the golden egg was laid by a cuckoo.

In his remarkably detached review of the evolution of the "federal grant universities," Clark Kerr declared that their growth has proceeded largely outside of "organized faculty influence and control." The principal influence, one must conclude, has been exerted by the flow of funds from the government. J. R. Pierce says that federal research support "is alienating engineering education from the civilian economy." Noting that "by far the greatest fraction of the total support comes from mission-oriented federal sources," he finds "this support is concentrated in two industries, electrical and communication equipment and aircraft and missiles." The effect of this bias in the flow of funds is decisive in the realm of graduate engineering education because ". . . the graduate students supply the universities with a new generation of professors as well as industry with some of its best-trained and intellectually most capable engineers." Under the circumstances, Pierce concludes, "It is no wonder that thesis subjects of Ph.D. candidates in engineering are strongly associated with the few sectors of industry relevant to defense and space. It is no wonder that, of those Ph.D. candidates who go to work in industry, a large proportion go to aerospace and defense industry, and that another large fraction who go to universities teach what they have learned in doing defense-supported and space-supported research." Engineering educators are not unaware of this problem, "but they have been subjected to inexorable financial forces, and these forces have established a pattern which makes it difficult for the educators to follow the dictates of their own wisdom."

Herbert York has made the shrewd surmise that it is engineers who create the shortage of engineers! Given the ambitious programs of technological enthusiasts in the Department of Defense and National Aeronautics and Space Administration, the committee on scientific manpower headed by E. R. Gilliland had no alternative but to project a continued soaring demand for talents trained in "EMP"—engineering, mathematics, and physics. York, himself, has concluded that there is no technical solution to the dilemma of "steadily increasing military power and steadily decreasing national security." Given a national policy committed to massive support of economic growth in the developing nations of the world, the Gilliland Committee might have found itself projecting a shortage of civil engineers, chemists, and agronomists.

A similar pattern of distortion by sheer weight of financing is to be observed in the expansion of research in the life sciences in the universities. The U.S. Public Health Service, beginning in the early 1950's, found itself charged with the disbursement of mounting appropriations for health research. Through the National Institutes of Health, it has come to support investigations at the most fundamental levels, remote from the practice of medicine. Because the agency has closest ties to the medical schools, most of its funds have gone to those schools. Only in the last few years has the percentage of appropriations going to other recipients, including the departments of biology in the faculties of arts and sciences, approached one-third. As a result a huge percentage of the frontier work in biology has moved over to the medical schools.

For the most part this work has no relevance to the task of professional education that occupies the regular medical faculty. Indeed, at a time when the behavioral sciences seem more relevant to the training of future physicians, the bestowal of a Nobel prize for work in the biochemistry of genetics upon a nonteaching research professor can demoralize students as well as faculty. At the same time, the undergraduates and the graduate students in the university "proper" are deprived of the presence of creative scientists who would find appropriate niches on the home campus for the teaching they are qualified to do. Some universities have tried to cure this situation by putting their biology departments and their medical schools under a single administration. So far such efforts have been defeated by the distinctly different interests of the two modes of education.

The mission-oriented programs have unquestionably sponsored great achievements in science. They also provide stipends, in one guise or another, for tens of thousands of graduate students. But the universities that have been the principal recipients of these funds show structural failure under the pressures they exert. Proponents of the project system have advertised the power of compulsion that might be vested in a single agency. They have overlooked the power of induction that is exerted by large concentrations of funds. It is clear, for example, that the character and structure of the engineering and medical schools have changed and evolved the way they have because that is the way the money was running from Washington.

Institutions outside the select circle of universities favored by the mission-oriented agencies have also felt the pressures generated by the fiscal flows. They have found it increasingly difficult, in the first place, to recruit faculty against the bidding for talent financed by federal funds. In the second place, they have little to offer in the way of research support to attract teachers of science. The National Science Foundation, to which they must look, managed in 1965 to grant half of its applicants a quarter of the total funds applied for by persuading the successful applicants to cut the scope or the term of their programs in half. The frontiers for the promotion of excellence are plainly wide open.

At this juncture, it appears, the delicate issues that surround federal support of the sciences are to be subsumed in the larger realm of issues, equally delicate, that surround federal support to higher education. No one can any longer doubt the mutually reinforcing relevance of the two activities to each other. The work of the scholar and scientist is only partly reflected in his published papers; the little or the much that he may contribute in the way of new knowledge comes out of his lifelong effort to comprehend and reconstruct the entire realm of understanding and ignorance that surrounds his research. This, the context of his work, is what he teaches. If he is successful his students join him in the framing of questions. The energy thus flows equally from student to teacher and drives the inquiry forward.

What is more, as Harvey Brooks has shown, "education provides the criterion for research support that is easiest to quantify." If America is now to fulfill its promise to provide every child with all the education he or she is willing and able to absorb, then higher education must become one of the most substantial growth sectors in the economy. The budget for basic research will have to grow along with it at a faster rate and to bigger dimensions than anyone has forecast on any other basis.

Questions of the gravest kind surround the impending massive flow of federal funds into higher education. The tone of public discussion has thus far been set by the great private universities. They have preferred to pretend that they have not already been receiving huge federal subsidies for their scientific activities. At one and the same time, their spokesmen have argued for higher overhead allowances on the ground that the research was undertaken at the government's behest, and they have justified the research as work their faculty would be doing anyway, in line with the high and proper function of the university. With overhead allowances now established in excess of 20 percent, these funds constitute a principal source of general income on the books of the biggest institutions. Nor does it advance the development of public policy to declare that the only alternative to the project system is the "formula," the "pork barrel," or "logrolling."

The design of the federal higher education program requires the responsible collaboration of the university community. Behind the distaste for politics and the fear of governmental control lie significant issues that must be faced in the open. It is not too early to begin the education of the electorate in the values of higher education and the central role of the university in the structure of our unwritten constitution.

No one can regret the continued rapid expansion of the undergraduate and graduate student population of the country. But it is clear that federal aid to higher education must go beyond scholarships and the building of dormitories and class-rooms. It will be necessary to invent new kinds of inputs to the capital assets of the higher education system.

The first object of such federal investment should be to fortify the universities as autonomous centers of creative initiative in the life of American society. Early in such a program, it would be desirable to seek to heal the damage done to our foremost institutions by their participation in the project-panel system of research support. Institutional grants for this purpose could well go to the strengthening of scientific departments that have been slighted by mission-oriented funds and the ebb and flow of fashion during the last twenty years. Federal funds can, of course, help to redress imbalance among the regions of the country, especially that caused by the neocolonial relations obtaining between the industrial-urban Northeast and the agri-cultural-extractive South and West. The pattern of regional interests may in the end prove less relevant than the pattern of urbanization; the wiser course, as urged by Lloyd Berkner, may be to build a strong university in each of the 100 major metro-politan centers of the country. In the deployment of such funds, there need be no inhibitions laid by the fading distinc-tion between public and private universities. Another major prospective demand for capital inputs is foreshadowed by the evolution of the liberal arts colleges into demiuniversities distin-guished by small and selectively specialized graduate schools.

As against the adoption of some simple-minded formula based upon population and income, in other words, the avail-ability of financing on the federal scale invites the most gener-ous and wise imaginations in American higher education to join in the framing of new objectives, new standards, and new kinds of fiscal instruments. The challenge grants of the Ford Foundation and the development grants of the National Science Foundation provide one working model. The project grant will continue to have a significant and creative role in providing funds responsive to "proposal pressure" and in stirring ventures into fields of learning that are neglected or that become important to the public welfare. Finally, the overhead allowance sets a precedent for federal contributions by formula to the unrestricted incomes of universities large and small.

There are undoubtedly many other lessons to be learned from the first venture of the federal government into higher education via its massive funding of scientific research. But the most important one is this: initiative and decision in the deployment of funds for research and education must be restored to the universities. Administrators and faculty members must insist that federal funds be extended on terms that respect the integrity and independence of their institutions. Those who celebrate "the participation of the scholar in the contemporary tasks and social movements of our society" speak with condescension of the "more inward-turning, conventional, and placid academic life." In fact, of course, solitude is the proper habitat of Man Thinking, and the quiet is evidence that the scholar is minding his own business. That business, as all of recent history shows, is more important than the busy business that disturbs the peace of our biggest campuses. Our universities deserve public support not as instruments of national purpose in the service of ends chosen by the government but as vessels that cherish and enlarge the liberties of self-governing citizens.

Peace and Quiet
on the Campus

THE merger of the Carnegie Institute of Technology and the Mellon Institute, a nonprofit industrial research laboratory, in 1968 created Carnegie-Mellon University, the twenty-first-largest privately endowed university in the United States. For the new university's first commencement that year I wrote the essay that follows.

I T IS more than an honor to be enrolled in the Class of 1968. To be counted in your company is a challenge. To rise as your commencement speaker takes considerable adrenalin. In our country and abroad, you and your classmates have asserted your presence in human affairs sooner and with larger impact than has any generation before you. On a long agenda of private questions and public issues, you have left no one in doubt about the warmth of your convictions: on the inexplicit premises of our sexual morality; on the objectives and the conduct of the war in Southeast Asia; on the comparative cost-benefit of alcohol and pot; on the vicious circle of race prejudice and poverty; on the propriety of military research in university laboratories and libraries. Your wisdom on some topics has paled before your sheer brisance. You have not only shut down a score of universities—from Madrid to Durham, North Carolina, from the Sorbonne to Columbia—you have brought down two national governments—though, by now, it is evident that we have not yet seen the last of Charles de Gaulle and we may yet hear again from Lyndon B. Johnson.

Now that you have joined me as fellow alumni, however, you must put aside student power. You will have to learn to work at domestic and world politics from the status of ordinary citizenship. As for those questions of personal morality, you will soon find yourselves speaking with the uncertain authority of parenthood. I do not presume to speak to you on these troubling matters today. As more befitting this occasion, I want to talk to you about the situation of your university and your obligations as alumni.

The first thing I ask you to consider is how much your university needs you. Carnegie-Mellon University is one of the thousand or so fully accredited degree-granting institutions of higher learning in America organized as private corporations. In its present, hyphenated incarnation it is the latest expression in the realm of higher learning of the distinguishing American genius for voluntary enterprise. With its public credit established by the names of two of America's great industrial fortunes, its existence can be taken for granted all too easily. Correspondingly difficult to believe is the proposition that the life of your university is on probation and that its future is in doubt. Not to impugn the credit of this university in particular, let me hasten to say that I am quoting, in effect, the words of John U. Monro on the occasion of his farewell last year as dean of Harvard College.

"It is easy," Monro said, "to take a big, functioning institution for granted. In fact, most big institutions like it that way and hire a high-priced image-maker to help us substitute a calm, impressive symbol for the difficult reality. If we are to be effective as citizens . . . we need a closer sense of the frailty as well as of the strength of our institutions."

The frailty of our universities, old and new, has been plainly

exposed in the academic year we are closing with this commencement. Things have been relatively quiet here, I take it.
But even here, in the atmosphere of single-minded determination that motivates professional education, there have been
stirrings. Elsewhere, on campus after campus, undergraduate
and even graduate-student innovations in extracurricular activity have made newspaper headlines and television spectaculars,
shocked the sensibilities of the community, and invited punitive
legislation in federal, state, and local council chambers.

Only a jejune conspiracy theory of history supports the charge
that these developments and disorders be laid to the Students
for a Democratic Society—however greatly events may have
fulfilled the expectations of this small, determined band. Nor
do frustration and despair at the prospect of compulsory and
unwilling service in a "dirty little war" explain the widespread
propensity for direct action. More significantly, I believe, the
generation now in our colleges has been studying the lessons,
not yet learned by their elders, of Nuremberg and Hiroshima.
At Hiroshima, it was our ordnance that placed the public safety
beyond defense by any power of the national state. It was our
lawyers who argued at Nuremberg that a man cannot delegate
the administration of his conscience to his superiors in the
chain of command.

In the first naked embrace of personal responsibility, students
have turned upon the institutions nearest at hand: their universities. But students also have a fresh and immediate apprecia-
tion of the central role of the university in the life of our
society. That role ought to be more widely appreciated among
the alumni.

The university closes the loop of power in a self-governing
society. It is, indeed, a function of the university to train young
men and women for service to society in occupations and professions requiring an ever higher order of intellectual equipment, to fill the intricate boxes in the manning tables of the
industrial system. The university has a prior obligation, however, to the student as the future sovereign of the republic. It is
the place where he undertakes his self-discovery, where he
makes the first moral and aesthetic choices that one day will
determine the goals and shape the ends of society.

"Students in every corner of the earth," Alexander Meikle-john wrote, "have the same basic lessons to learn. They need to know each other. They must become acquainted with that whole human undertaking which we sum up in the phrase 'the attempt at civilization.' Only by having that common knowl-edge can they become reasonable in their relations to one another. 'The proper study of mankind is Man.' "

Again, our universities have assumed well-advertised func-tions as centers for "problem-solving in the national interest." They have a prior commitment, however, to inquiry that may redefine the national interest. The question "What is science?" is reserved to the university not only because it is of interest to philosophers but because it lies at the very crux of the scientific enterprise.

As Willard V. Quine shrewdly observed, in the course of settling that philosophical question:

It would be unwarranted rationalism to suppose that we can stake out the business of science in advance of pursuing science and ar-riving at a certain body of scientific theory. Thus consider the smaller task of staking out the business of chemistry. Having got on with chemistry, we can describe it *ex post facto* as the study of combining atoms and molecules. But no such clear-cut delimitation of the business of chemistry was possible until that business was already in large measure done. It is a commonplace predicament to be unable to formulate a task until half done with it.

That is why the university scientist may not work at any science but his own. The asking of questions—the formulation of prob-lems, not the mere solving of them—is the proper function of the university.

To enlarge human understanding and to teach the young, it is apparent, the university must be free. The autonomy of this institution must be hedged by the same sovereignty that secures the liberty of the citizen, for it is the citadel from which the citizen deploys on the perilous missions of inquiry and conse-quent social change. If students are disquieted about the inde-pendence and integrity of their universities, then surely that disquiet merits the consideration of other interested parties.

A candid appraisal of the situation must begin with the

acknowledgment that the American university is, by constitution, weakly designed to carry its heavy responsibilities. It is a corporation on the standard domestic model, with owners (the trustees), managers (the administration), and employees (the faculty!). Now, the corporation is a useful instrument for getting things done, as is demonstrated by almost everything that gets done in this country. But it is not a democratic institution. I know, because I am president of one, myself, and I would not have it any other way.

The status of employee is, by definition, inimical to the liberty of the scholar. Yet the corporate vehicle has been successfully adapted to contain and foster the university enterprise in our country. That adaptation has been accomplished principally by the faculty, organizing itself as a community for the purpose of mutual protection and to secure tenure—"appointment without limit of time"—as the guarantee of the independence of each scholar. Thus, in our university corporations, trustees and administrators have learned to be content and proud to serve as the providers and conservators of the resources that sustain the inquiry of the scholar and the learning of the student.

This pragmatic arrangement has come under increasing stress during the past twenty years, and the university structure is showing signs of increasing strain. To begin with, the number of students enrolled in colleges and universities has more than doubled in this period. One-third of the country's college-age cohorts are going to college. Most of the increase in population has been absorbed by public institutions, through expansion of enrollment on existing campuses, the building of new campuses, and the creation of new state college and community college systems. Crowded classrooms, intellectually undernourished students, overworked professors, and administrations not yet secure in their relation to public authority have made these campuses into places of turmoil and unrest. We cannot yet claim that our rich society has made good on the founding fathers' underwriting of the perfectibility of man.

There is another index that plots the rising stress and strain as faithfully as the enrollment curve. This is the increasing flow of federal funds to higher education. The first stress exerted by federal funding arises from the fact that nearly all of it goes to science. Plainly evident in the structure of the major universities is the strain that divides the faculty of arts and sciences. The law of *pars inter pares* that, in principle, binds the community of scholars is breached by a disparity in income. The gross economic discrimination may be taken as a measure of less quantifiable but no less perceptible divisions of interest, intellectual commitment, loyalty, and lifeways that have split the faculty of arts and sciences down the middle. The mutual repulsion of scientists and humanists promotes a downgrading of concern for values on the one hand and a tide of anti-intellectualism on the other.

Other strain lines run in many different directions through the faculties of science. The stresses here reflect the nature of the contractual relationship between the university and the government and the practices of the federal granting agencies. Virtually all of the funds flow in support of research projects upon application by individual scientists. This practice is celebrated as placing emphasis upon the merit of the work; it is pointed out that nearly all the money, again, is disbursed on recommendation of the investigator's peers in his particular line of work.

Project funding plainly exerts disruptive pressures on the structure of the university. It detaches the scientist's loyalty from his faculty and transfers it to the "invisible college" of his field. It has brought the growth of this and then that department principally in consequence of the changing pattern of the availability of federal funds and not in any necessary accord with university plan or policy. Project funding has also, on occasion, invited the kind of enterprise that prompted James Perkins to observe: "It is the casual, unreflective opportunistic development of interests for the sole purpose of attracting funds for prestige which obviously violates integrity."

Until the last few months, scientists and administrators deprecated any suggestion that science in the university has been placed under untoward ulterior pressures arising from the fact that mission-oriented agencies supply 90 percent of the federal funding. Lee DuBridge faced the issue with the declaration that the federal government indeed "purchases" research from the universities for such objectives as "our national defense, our prestige among nations, the health and comfort of our people, the viability of our economy." To this DuBridge stoutly adds, the government has been " 'purchasing' from the universities the basic research they wanted to do, rather than tasks specified by the government." The support of science as a byproduct of "problem solving in the national interest" has constituted an arrangement agreeable to all concerned. It has been ratified by every committee on the question appointed by the National Academy of Sciences; in particular, it has been found preferable to any attempt to channel flows in response to a national science policy.

With funds running in abundance, especially to the largest universities, from the Department of Defense, the Atomic Energy Commission, and the National Institutes of Health, no one was disposed to urge larger appropriations for the National Science Foundation, the one agency charged with support of university science. Intramural stress and strain at the university were relieved by the constant increase in funding, proceeding at times by as much as 20 percent in a year. Imbalance or inequity in the distribution of this year's grants could be redressed with a little more grantsmanship by the slighted parties next year.

Suddenly, the horizon has closed in on this endless frontier. It is not that the money has stopped. There has been a retardation in the rate of increase. This follows predictably from the absurd result of the extrapolation of the earlier slope that showed the national "R and D" budget exceeding the gross national product by the year 2000. In actual fact, the retardation is probably no more than premonitory. It is forced by the budgetary constraints of the moment. Very likely, the budget for science will be returned once more to that earlier slope before it is brought into a more stable and considered relationship to the growth and structure of the economy.

The present constraint affects principally the flows from the mission-oriented agencies. The valving down of NIH funds for the life sciences has produced a crisis in medical education, manifested at its extreme by the threatened insolvency of four medical schools that have depended upon research grants to finance their teaching. Physics feels the pinch next most severely, with the phasing out of new starts on strategic weapons systems and the slowdown of the space adventure. The Physics Survey Committee of the National Academy of Sciences found in 1966 that the increase in federal funding—down from 16 percent to 6 percent per annum—was then barely keeping pace with the rising cost of doing physics.

Meanwhile, it turns out, graduate enrollment in physics has been declining relative to the growth in graduate education, and, more ominously, undergraduate and high school enrollments have been declining absolutely. The Physics Survey Committee attributed this development in major part to the shortage of physics teachers created by the overwhelming competitive demand for physicists in the giant enterprises of applied physics. Harvey Brooks offers the further explanation:

Science—and, above all, physics—has become inseparable in the public mind from the successive revolutionary changes in military technology and from nationalist competition in space spectaculars. . . . This identification which undoubtedly benefited basic science greatly during the period from Sputnik (1957) to about 1963, has increasingly reacted to its disadvantage, especially among the generation that does not remember World War II.

That generation, Brooks adds, regards the present relation between science and the federal government as a "nefarious alliance"; this alliance furnishes "an underlying theme of the student protest movement."

The time has surely come to call in question the strategy or nonstrategy that ties the support of science in the university to the fortunes of the mission-oriented agencies. Apart from the now manifest unreliability of this kind of funding, it has helped to confirm in the public mind a misunderstanding of the true nature of science and of the role of the university in our society. The mission-oriented agencies must justify the "mission-relatedness" of their research expenditures. Rationalizations to this end have been generously supplied by the parties of the second part in the writing of research contracts and of policy statements addressed to wider readership. As Harvey Brooks has recently observed: "One of the unfortunate side effects of demanding too immediate a pay-off from a basic research field is that it tends to lead the practitioner into a species of intellectual dishonesty not only with respect to the public and the politicians but even in their own minds." In consequence of this kind of salesmanship over the past twenty years, science has established no substantial claim of its own on public support. Worse yet, some practitioners have brought into the universities problem-solving projects that bear little connection to the mission of inquiry and none to teaching. As long as ten years ago, a president of this university, J. C. Warner, called for "Peace and quiet on the campus"!

No useful purpose can be served by continuing to isolate the funding of science from the difficult and novel questions surrounding the relations of the federal government to the university as a whole. At 20 percent of the budget of the accredited universities, the federal funding must be recognized as a substantial subsidy to their total operations. At 40 percent of the budget of our foremost universities, federal funding surely engages the structure and integrity of these entire institutions. Every circumstance argues that the position of government in the university system is sure to grow and not diminish in the future.

University endowments add up to a grand total of only $12 billion; these funds yield much less by way of current income than is yielded by project contracts and grants from Washington. The bravest resource-building enterprises of the richest universities cannot reverse this relationship of public to private financing.

The first task of university administrations today is to reconstruct the fiscal arrangements that tie their institutions to the federal government. Significantly, this statement applies as validly to the great state universities as to the private in the list of the top 25, 60, or 166 institutions in the country. The primary issue to be defined and negotiated is not "how much"; the issue is rather "on what terms" the university can properly solicit and accept federal and other public support.

The interest of the university can be clearly stated, especially on the rebound from present arrangements with federal granting agencies. The terms of support must respect the self-governing autonomy of the institution and the self-motivating freedom of the scholar. Public policy must declare funds available to the support of education and free inquiry as significant social enterprises in themselves without compromise by ulterior motives. Commitments of funds must be made for long periods of time. Major grants should go without restrictions to the general income of the university and so to the restoration of the faculty community.

There has been some timid experimentation with "institutional" grants, largely by the National Science Foundation out of funds for support of research that are regularly four times oversubscribed. The present ratio of institutional to project funds ought to be reversed. Project funds would have their place in the new arrangement; with the independence of the university scientist secured by funds appropriated to the support of science, the mission-oriented agencies could be permitted to write project contracts. What is implied here, however, is substantial increase in the funds running to the universities from the National Science Foundation, from the National Endowments for the Arts and the Humanities, and from the Office of Education. As against the creation of a department in the Cabinet, it would be well to maintain a plurality of granting agencies, each concerned with the cultivation of its territory, and to structure the total federal flow through a National Council on Education, Science, and Culture, comparable in authority and stature to the National Security Council. The President's Science Advisory Committee and the review panels of the granting agencies commend themselves as models for the channels of communication needed to bring the university constituency into the making of policy.

Pleasant as it is to block out the design of Utopia, it is more important now to recognize that higher learning in America is already involved and is bound to become more deeply involved in the political process. With 10 million college graduates at large in the population, the university community ought to be less intimidated by the thought of facing the electorate. Each university will continue to depend upon its own alumni constituency within the electorate to promote its interests and defend its integrity.

The exponents of student power thus need not, upon advancing to the grade of alumni, abandon their concern to promote the perfection of their universities. They can join their alumni associations! Nor will public funding deprive these latest alumni of the opportunity to contribute to the financial support of their universities. On the contrary, each alumnus will be moved to more generous outlays by consideration of the leverage his dollar exerts upon the flow of public funds.

VII Education for Self-Government

We Good
Americans

THROUGH the academic year 1970–1971, the campuses have been quiet. It may be that the student rebellion that began in Berkeley in 1964 has run its course. The post-mortems, now coming in, ascribe the malaise of the young variously to surfeit, to decay of the middle-class home and values, to Oedipus complex and joyless nurseries. There is much else in the American scene that fits such diagnoses. If that were all, the waning of the rebellion, with its business unfinished, would not need regretting.

That there was something more, and more serious, on the agenda was strongly indicated by the first of the autopsies, the report of a faculty committee set up in 1965 to consider educa-tion at Berkeley in the aftermath of the Sproul Hall bust. Among its findings, the committee reported: "Students tend to fall into two distinct categories as regards their support or opposition to FSM [the free speech movement]. . . . The [academic] major with the highest percentage of its students arrested was the Social Science Field Major (14 percent). The next five majors, all with more than 10 percent . . . were in the humanities." On the other hand the majors with the lowest or with no representation on the police blotters were the vocational and professional fields of agriculture, forestry, optometry, and public health. "The sciences fell in the middle, with the biological sciences having a higher proportion of arrested students than the physical sciences." A footnote records the fact that "the major with the highest percentage was actually Molecular Biology, with three students out of 14 arrested or 21 percent."

"The students arrested at Sproul Hall," the report goes on, "also included an unusual percentage of scholastically able young people. Their grades were significantly higher than those of the average student. Nearly half had grade-point averages higher than 3.0, whereas only 21 percent of the total student body had grades this high." The same bias appeared among supporters as well as participants: "Over four fifths of the students surveyed in April 1965 said they agreed with the goals of FSM, although only one half approved of its tactics . . . 80 percent of the surveyed undergraduates who had grade point averages over 3.5 approved of FSM, but only 44 percent of those with less than 2.5 grade point average did so."

The committee concluded:

The experience of FSM offers serious lessons. First of all, the ease with which a majority of the students could find, however ephemerally, a commitment and a moral drive in opposing the University administration is evidence of a widespread, if latent, alienation. . . . Secondly, the high intellectual abilities of many strongly committed members of the FSM may mark them for positions of leadership in our society, particularly in cultural and political fields. The success of the movement demonstrated their capacity for leadership.

Their fellow citizens cannot afford not to take seriously the issues that engage such young people. Considering further, that the twentieth-century children's crusade has seized countries in Europe and Asia, most notably in France and Japan, and has mobilized the young in socialist as well as capitalist countries, the spur to action by American youth cannot be put down to exclusively domestic and parochial stimuli.

The one common cause that reaches across all national

boundaries is the common destiny portended by the Bomb. Having completed my education and formative years long before Alamogordo and Hiroshima, I do not pretend to know what it has been like to grow up in Its presence. I know that it has taken considerable conscious and subconscious dulling of my sensibilities to work and find joy in a world whose recent history has proved no less absurd than my first rush to hope that the attainment of the ultimate weapon could compel world peace through world law. In our country, to be sure, the sense of foreboding is deepened by engagement in the ugly and futile war in Southeast Asia. For the young men who face drafting to engage in unnatural acts of personal violence, that war throws a lurid light on all questions.

If physics has confronted society with an ethical dilemma, the young comprehend that the biological sciences pose questions of a more direct and personally ethical nature. Leo Szilard, author of the letter to Franklin D. Roosevelt that initiated the Manhattan Project, said he was cheered about the prospects for human survival because the species was surviving the choice offered by contraception. Women's lives are no longer committed to childbearing, and childrearing is no longer the perpetual and sufficient motivation for their existence. Relations between the sexes need not now involve consideration of either an immanent or an imminent third party. Student power has compelled recognition of this development in the living arrangements and mores of the campus.

Much as parents and other adults may recoil in anxiety and dismay from such dismantling of the sexual code, the young apprehend still more profound ethical questions in biology, questions to be faced by them now as they form their values. It is possible, for example, to transfer the nucleus of a tissue cell to an egg cell from which the nucleus has been removed. From this mechanically hybrid cell an exact replica of the donor of the tissue cell arises. Somehow the cytoplasm of the egg reawakens the full genetic competence of the tissue-cell nucleus. Since all of the tissue cells of an organism carry the exact same heredity, here is a way—prospectively—to produce an entire regiment of replicas of a single individual. Biology, extending the systematic confrontation of reason and experience to the identity of man himself, has completed the demolition of the foundations of received value begun by celestial mechanics.

The traditions of higher education in our civilization say that the young who are fortunate enough to have its advantages should pursue their concerns by rational inquiry and civil discourse. A tide of anti-intellectualism runs strong, however, in the universities. The radical attack proceeds in many quarters from conviction in the determinism of historical processes deriving from Marx and, at the same time, from commitment to personal moral responsibility for history that comes from Sartre. In reason, the two approaches are mutually exclusive. Yet, mutually exclusive as they are, they set up a fulminating explosive drive. Fatalism and absolutism made formidable adversaries of the students who have manned the barricades.

If the thought processes of this college generation are so alien to the rational tradition on which our civilization is founded, then surely their schools have failed them. To the nature of education for self-government the three essays in this section are addressed. My reflections here unavoidably engage the subversion of education and self-government by our country's brutalizing venture in Vietnam. In the first of these essays, addressed to the students at Phillips Academy on Moratorium Day in October, 1969, I did not reckon with Vietnamization cum mechanization of our continued military commitment, on which the Nixon administration now appears to have placed its political bets. If the recall of the draftees from the ground there—leaving our Air Force and Navy to supply the South Vietnamese troops with firepower—does succeed in defusing opposition to the war at home, then the scar tissue may seal in a fatal abscess.

ANTHROPOLOGISTS have a name for a gathering like this. It is a "rite of intensification." We have nothing new to say to one another. We are gathered to repeat once more the truths, the obvious and the terrible truths, we have been telling one another, month after month, throughout the course of our tragic adventure into Southeast Asia.

To begin with, we all agree and declare: "We want peace!"

Yet in my lifetime, which seems short enough, our country has fought four major wars to keep the peace. How explain this schizophrenia? We declare simple truths in plain words, and then we deny them with all the massive destructive power at the command of our society.

The cause of our confusion is to be found, perhaps, in the still incomplete progress of mankind toward civil order. Within our national communities, we establish the rule of law. But the rule of law stops at the boundary of each national state. Among the states, anarchy and the law of the jungle prevail. We see on the other side of national boundaries, when we look in certain directions, not fellow-men but members of this or that "pseudospecies"—a powerful word and idea we owe to Erik Erikson. So it is possible for us to declare that war is a monstrous and inhumane enterprise and, at the same time, to foster the making of war as an honorable profession.

It may be that mankind is in transition to a new international order in our time. We find security in the insane moral contradiction which declares that weapons made for mass destruction have made war unthinkable. The threat of species suicide imposes what order now prevails among the national states. Yet we continue in our old ways to court the thermonuclear catastrophe. History proceeds on its violent course in the illusion that there are now two kinds of war. Not only the big industrial nations but even the new nations emerging on the old colonial continents think they are free, under the cover of the stalemate of thermonuclear terror, to fight limited war by conventional weapons.

Our experience in Vietnam shows this to be an increasingly dangerous game. The two kinds of war are becoming indistinguishable.

In Vietnam our country has expended a tonnage of violence that exceeds all of that poured into the European theater in World War II. The landscape from the Mekong Delta to the highlands at the DMZ has come to resemble, over vast stretches, the surface of the moon and of the equally lifeless planet Mars. In the cascade of violence we have introduced entirely novel refinements. Puff the Magic Dragon is a helicopter that carries the fire power of three machine-gun companies; the hail of steel from this weapon has been observed, in operations over the suburbs of Saigon, to fall impartially upon the just and the unjust, upon friend as upon foe. A so-called tear gas is being employed on a large scale to circumvent the frustrating tactical doctrine which declares high explosives to be ineffective against personnel; this is more than a tear gas, an asphyxiant that drives the target from his foxhole to expose him more nakedly to bullets and schrapnel. We are using another biological weapon in the defoliants that have stripped the vegetation from whole landscapes in order to interdict the foe's food supplies, with ecological consequences still to be reckoned.

For all of the expenditure of violence there are no battles won, no territory taken. There is nothing to celebrate but the daily body count. The dreary numbers—the dubious scorekeeping that asks Americans to find satisfaction in the report that our young men have killed more of theirs than they have killed of ours—are haunted by the legions of dead Vietnamese civilians whose bodies do not count.

The record of impersonal violence is now highlighted by the revelation of acts of individual and personal barbarity committed by American officers and men. In the new tactical doctrine of counterinsurgency, they have pioneered modes of torture and murder not contemplated in the rules of war. They and we—good Americans at home—must live the rest of our lives with the guilt of Nuremberg in our hearts. For it was our lawyers and judges who propounded at Nuremberg the doctrine of individual moral responsibility up and down the whole chain of command, the doctrine of the complicity of the people in the crimes of their leaders.

After six years, this war that all agree is unpopular, expensive, and unwise has become, for a mounting number of Americans, a war that is immoral, inhumane, and intolerable. The prolongation of this agony has divided our country with a bitterness not known in a hundred years. The neglect of domestic priorities for the expenditure of our material wealth in violence overseas has poisoned our domestic politics. The demand upon our young men to make unlimited sacrifice for limited objectives—to face death for the political convenience of two national administrations that have been unable to manage the political embarrassment of withdrawal from this fruitless war—has now disrupted the moral community of our universities. Politics in our country has moved "into the streets," into modes of protest unknown to our political tradition.

There are some who say that this Moratorium Day has been called against the national interest. It will ruin, they say, the bargaining position of our country. A former Secretary of State has declared that the opponents of the war are recklessly engaged in an attempt "to break another President." The Vice President declares that this day's demonstrations have been called to bring aid and comfort to the enemy.

On the contrary, this day's message is addressed not abroad but to our country, and not against but to our President. It declares that Americans will be content with whatever deal Washington can make to secure the withdrawal of American forces from Vietnam at the earliest possible date. This day declares that the administration will find no political embarrassment in charges that they have taken a defeat. President Nixon and his party need not fear the charge which Senator Nixon laid to the Truman administration in "the loss of China."

With the free hand thus extended to him, with his domestic flanks and rear cleared for maneuver, it is a graceless political humor that prompts the President's spokesman to invite the opposition to negotiate the settlement in Vietnam. The opposition has made its contribution, in fact, to the thinking process. From Senator Charles E. Goodell has come the proposal that American troops be withdrawn before the end of 1970. Such action would compel Saigon to assume the responsibility it asserts and the legitimacy it claims as the government of the people of South Vietnam. From Cyrus Vance, who conducted the first ten months of negotiation with Hanoi in Paris, comes the proposal of a "cease-fire in place." For both sides negotiations would then appear preferable to renewed warfare. Good ideas, it appears, are not enough. This day's demonstrations across the nation are required to press upon the administration the urgent task of thinking its way into action for peace in Vietnam.

In default of such enterprise from the White House, the making of foreign policy has devolved upon the Pentagon. For five years the generals have made the only kind of policy they are qualified to make. The civilian electors now insist that their Commander in Chief reclaim his constitutional authority over this mindless extension of diplomacy in Southeast Asia.

The Tet offensive, in the first weeks of 1968, demonstrated another truth that requires repetition here: The war in Vietnam cannot be won. Military victory is not open to either side. There can be no victory for the Vietcong and North Vietnam so long as we keep our army in the field. But there can be no victory for our arms against their apparently inexhaustible willingness to die.

For all of our expenditure of violence and counting even the bodies not counted by the Pentagon, our military exertions have scored fewer casualties in the tiny territory of South Vietnam than World War II in all of Europe. This is a measure of the toughness of the tissue of the civilization of preindustrial agriculture, with its villages spaced out like gopher holes, as compared to the fragility of modern industrial urban civilization. We may hope that the stubborn persistence of our enemy is a measure of still more universal qualities in the human spirit, qualities in which we ourselves may have a share. The 2,000-year history of Indochina shows we are only the latest foreign power to feel the indomitable will of these people to be free.

It is plain, by now, that we are not in Vietnam defending a free nation against external subversion and invasion. Nor can our presence there be described as intervention in a civil war. For thirty years before we came on the scene, the indigenous Vietnamese leadership had been pitted in rebellion against puppet tyrants installed and sustained by France, by the Japanese, and then again by the French. We have succeeded the

Chinese, the French, and the Japanese as the imperial occupying power. This was surely not our explicit intention; our position in Vietnam was taken with different ideological motives and larger strategic objectives in view. In the eyes of the Vietnamese, however, we are the old, familiar enemy. The Vietcong and the troops from North Vietnam could not possibly maintain military power sufficient to pin down 500,000 U.S. troops without the cover provided by a population friendly to them and hostile to us. The weapons count is less well publicized than the body count. When the numbers have been uncovered on occasion, however, they show that the United States is the major supplier of arms to the enemy—the weapons finding their way through channels provided by our thoroughly corrupt and often disaffected Vietnamese allies.

The military scales are thus weighted heavily against us by another truth that scarcely calls for repetition: We have long since been politically defeated in Vietnam. The Thieu government is the fifth generation of satrap we have propped up in Saigon. It is illuminating to recall that we started with a hereditary prince, Bao Dai, whose legitimacy had been supplied by the French and cherished in turn by the Japanese. We installed the Diem family in his place; we benignly condoned their extermination by General Khan and then suffered his replacement by Ky, now biding his time in eclipse by Thieu. Needless to say, the ultimate destination for the actors in this shabby procession, those whom it does not first tumble into the grave, is the Côte d'Azur. In Hanoi, by contrast, we face the national leaders who fought the Japanese and defeated the French at Dienbienphu. Our intervention on the side of Bao Dai cheated Hanoi of the fruits of that victory: the national plebiscite guaranteed by the Geneva Convention of 1954 that promised to unite North and South Vietnam under a single government.

This rehearsal of familiar truths brings us at last to repeat the antecedent truth: Our continued presence in Vietnam serves no national interest. We have no perceptible economic stake in the meager resources of the country. The domino theory that followed from the original, primitive, simplistic, bipolar cold-war model of the world polity has long since yielded to the plural-ism that has divided the socialist side, as well as our own, on more traditional and deeply rooted lines of national interest. Only our intervention in Vietnam could have brought that country as dangerously close to the gravitational field of Peking as it appears today. If Owen Lattimore had indeed been "the architect of our foreign policy in the Far East," the 2,000-year resistance of Indochina might have been enlisted to buffer, with far more stability, the southern flank of the emerging world power of China.

Nor does any obligation to "a free government" or even to "friends" tie us to Saigon. For the past two years, our beneficiaries in Saigon society have been crowding the outbound air transports for Europe, their liquid capital preceding them to Switzerland. For those that get caught in the hastening debacle, it ought to be possible to arrange, as we did for the refugees from Havana, a Miami somewhere in the Far East—at Macao, perhaps—where they can continue in their accustomed trades.

So, we can take the President's word for it and join in reaffirming that: We are getting out of Vietnam. From his declaration of this intent, we dissent only as to the timetable. In the attempt to salvage the face and front of the U.S. command, the President proposes to phase our troops out of Vietnam as Saigon demonstrates its capacity to "Vietnamize" the war—to last long enough to go under after we have gone. All the evidence shows, however, that the war is already Vietnamized—with the people of Vietnam increasingly united against us. Our Saigon friends are already getting out, discounting by months in advance the announced pace of our withdrawal. This Moratorium assures the President that the U.S. electorate will accept the withdrawal of our forces at a still faster pace. The worst outcome of the present accelerating trend would leave our troops to fight their way out alone to the air and seaports, with not even the fiction of a South Vietnam government to cover their presence in the country.

The lesson of Vietnam, we hear it said, is that our country must now stop trying to play policeman to the world. We have paid enough in treasure, blood, and honor to learn a more constructive lesson. The peoples of the old colonial world are committed to revolution, to political, social, and industrial revolution. Thus far, at every point, from the Bay of Pigs to the Tonkin Gulf, American power has placed itself in the way of this irreversible onrush of history. Vietnam shows we cannot stop history. We must turn to the redefinition of our national interest. To this end we ought to rediscover the American history that still lights the beacon for so many of the world's striving peoples. We could prepare for a more hopeful role in the world revolution if we would secure at home, to all Americans, the promise of equity, justice, and liberty held out by our own revolution.

The Imaginative Consideration of Learning

IN WHAT I have to say in the essay that follows, I owe a debt to two great teachers, Alfred North Whitehead and Alexander Meiklejohn, already emeritus when I came to know them. This essay was addressed to a gathering of faculty, students, and alumni at Phillips Academy assembled in April, 1970, to consider the school's future, starting with the question whether this oldest of the New England academies still had one.

THE high school years are surely the interesting years in education, for they come at the moral-ethical crisis in human development. Erik Erikson has reminded us that the prolonged infancy of our species is the foundation of civilization and culture. He has shown us also how mismanagement of that prolonged infancy is the source of all that is inhuman and destructive in man. The great task of education, especially in these years, is the freeing-up of the human spirit to take on the world. On the Erikson model, the young at this point in their lives are in transition from the state of moral being—that is, conforming out of love or fear to externally imposed codes of behavior and morality—to the adult stage of ethical being—that is, able to act in accordance with their own programs of hope and aspiration.

To this turning point in the growth of the young what does Phillips Academy contribute? When we ask questions and peel back the layers of the proverbial onion, we come to the last question. We finally must ask: What is Phillips Academy doing here at Andover, in the 1970's? By what does it justify its existence?

The Andover that each of us cherishes is our personal model, made up out of our experience here at school. That is necessarily an Andover that is already fading into the past. The longer we have been out of school, the more distant and wraith-like that vision must be. In other words, nostalgia is not enough. We are talking about the future. If the school is not serving the function that it served when we came here, then what function should and can it serve?

This is an important question to us as citizens as well as alumni. Phillips Academy is among the hundred richest private enterprises in American education, richer than such famous colleges as Colby, Hamilton, or Lawrence. As I have come to know the inside workings of the school, moreover, I see how inadequate its resources are to the needs and the claims that crowd it from without and within. We have the obligation to make those resources serve effectively the best uses we can choose for them.

When most of the alumni here present came to Andover, this was a preparatory school that equipped us for admission to the college of our choice. The sons of the American meritocracy whose families could afford to send them here were lucky to find that Andover was able also to leaven their community with the presence of young men on scholarships. Better than any local high school we were likely to go to, Andover prepared us for the college board exams. The strict, high standards of the school supplied what we required in the way of "cognitive" learning, as it is called today. The school of those days also contributed to something called "affective" learning. It got us away from home and gave us a conditional independence, preparatory to the independence we were to find at college.

Today the place for that kind of school in America is vanishingly small. In the 1930's the median education attained by white males—as the United States Census phrase goes—was a little more than eight years; no more than about 10 percent of the high school graduates went on to college. In 1970 the median education attained by white males is twelve years plus, and something on the order of 40 percent of the college-age cohorts are going on to college. The enrollment of independent boarding schools like Andover was always a small percentage of the total teen-age population. These schools educated, however, a significant percentage of those who went to college in our day, especially to the Ivy League colleges. The prep school graduate was the archetypic figure in the undergraduate communities of those colleges and strongly conditioned the style of life there. Through the important role they played in the Ivy League colleges, the prep schools exerted a leverage on American education out of scale with their actual size. This period that has seen the median education attained go up by 50 percent has also seen America respond to the promise of its eighteenth-century founders by making higher education available to nearly half the young people of college age. As the high school boy population increased from 4.5 million to more than 7 million, the number of preparatory boarding schools did not increase. Nor did applications to the existing prep schools increase with the rise in the high school population.

The famous New England prep schools find themselves, in consequence, educating a negligible percentage of the boys in high school. In the university colleges of the Ivy League, the prep is just one of a number of an increasing diversity of young men and—even—women. As these institutions have sought identity as national universities, they have correspondingly democratized and diversified their undergraduate communities. An Andover diploma no longer, therefore, guarantees admission to the college of your choice. The prep school—this school—has less leverage and influence. The leavening it may work in the system is less perceptible, by a good deal, than it was.

Suburbia—the new America of our time—boasts dozens of high schools able to challenge the prep schools for places on the admission lists of the colleges. The economic sacrifice for tuition and board and the rest has become a forbidding hurdle to family incomes after taxes and heavily constrains the motivation to send a son away to school. And, when it comes to affective education, the young men who might otherwise want to go away to school are more likely to find—or to think they will find—more of what they call "freedom" at home than cooped up in some boarding school. That four-letter word "girl" has a decisive place in the future plans and expectations of Andover and other prep schools.

Here it ought to be mentioned, parenthetically, that the well-off people of the world are growing up faster and arriving at biological maturity earlier. The age of arrival at puberty in the industrial societies has gone down by two years since the beginning of this century. This hastening of biological development now challenges the pace and scope of the prolonged infancy of education.

All of these developments sum up in a considerable economic pressure on this and other boarding schools. Applications have not only been declining as a percentage of the total population; they have been declining absolutely. The raw compulsion of the market compels reexamination of the mission of this school.

There are other developments that throw doubt on its mission. When young men go from this school to the college of their choice, usually one of the university colleges, they discover it is not a college but another prep school. It is a prep school for a graduate school—in the arts and sciences or in one of the learned professions that is practiced outside of the university. The professionalization of learning reaches back from the graduate schools through the university college to the very first courses in whatever course of study the student proposes to take. Those primary courses that ought to provide a speaking acquaintanceship with a new realm of knowledge turn out to be just the first step on a ladder that is supposed to lead, in seven or eight years, to a doctoral degree in the subject. The back pressure from the graduate degree governs and confines, channels and narrows, the character of the educational offering across the boards in the colleges. There are many points of entry to careers that diverge in many different directions, but there is no demonstration of the unity of knowledge, no way to get a look at the horizon.

This trend in the evolution of the American university has proceeded to the point where the University of California, which has led the nation in the popularization of higher education, is actively considering the exclusion of the first two years of college from its major campuses. It would turn over the "liberal" phase of education to the state college system and the newer two-year community college system that has grown up alongside of and independent of the university. In the great private universities of the East, there is a school of educators who argue that the university should devote itself to graduate education, beginning perhaps with the preprofessional concentration in a field in the last two years of college. One persuasive spokesman for this proposition is Carl Kaysen, who left Harvard a year or so ago to become head of the Institute for Advanced Study at Princeton, a center of learning that has no students at all.

To bring this development closer to home, let me tell you that there is on its way into print the report of an enterprise called "The Four School Study," a study conducted by a member each of the faculties of Andover, Exeter, Lawrenceville, and Hill. Considering that the study was sponsored and supported by the Carnegie Corporation, it is a sort of royal commission report. The Four School Study faces some of the questions I have put forward here. It concludes that these four schools, and other prep schools that hope to survive, will be able to do so only as they assume the role of intermediate colleges. Their function would be to prepare their students for entry into the university colleges for preprofessional training in what would be, chronologically, the last two or three years of college. In place of the cycle of eight years of elementary school, four years of high school, and four years of college, for example, the report envisions a cycle of seven years of elementary school, three years of junior high, three years of senior high or intermediate college, and three years in the university, leading to a preprofessional master's degree in the student's chosen speciality.

It is thought that Andover might become an intermediate college embracing the last two high school years and the college freshman year. If Andover did so, it would undoubtedly send on most of its sons to the universities for their preprofessional M.A.'s. For many, however, it would provide "terminal education," at the chronological age of the present college freshman but at an educational attainment equivalent to that of the sophomore year. Andover and other prep schools would thus, presumably, relieve the great universities of the burden and obligation of "liberal" education.

This sets out a sufficiently revolutionary development for our time. We ought to recall, however, that this was the sort of education Andover offered in the nineteenth century. Consider those magnificent young men posed like shades of Elysium in the photographs of the Andover athletic teams of the 1870's that adorn the gymnasium. In their abundant beards and mustaches, they look like college seniors. The innocent faces we find under equally impressive whiskers around here today suggest that they were just as young and unlettered as the contemporary prep school boy. For many of those young men, however, Andover did provide terminal education. Phillips Academy was still something like the original "New England academy."

Now, I have questions, not answers. In particular, I have serious questions about this prescription. It brings to mind a parable told by the Viennese architect Frederick Kiesler. Long ago, before World War II, it seems, the city fathers of some town in Germany were confronted with the need to do some rebuilding of their zoo. They mounted what we would call today a "total system study" of zoo life. From inspection of other zoos around the world the animal behaviorists on the team came home with a pitiful finding: everywhere the poor animals were housed in square cages, and everywhere they walked in circles inside those cages. To the designers and the architects the lesson was clear: Build round cages! So now, even to this very day, you can see the animals in that zoo walking in squares inside their round cages.

The Four School Study, I fear, prescribes round cages. My suspicion is heightened by the knowledge that the prescription issues from the "multiversity." This is the American university turned service station; the institution that boasts of its administration in troika with its faculty and students; the giant contractor to the defense and space and any other agency supplied with currency; the "instrument of national purpose" engaged not alone in the familiar missions of education and research but in the rewarding new mission of service to public policy.

In the years since the end of World War II, the service mission has produced directly and indirectly from one-third to 85 percent of the total income of our biggest and grandest universities. It has financed, beyond doubt, a triumphant epoch in American science. But this development has had other consequences that reach to the very essence of what a university is supposed to be.

I shall mention just two essences and consequences. A university is supposed to be a source of change and innovation, of external criticism and detached judgment of public policy. As an agency serving policy it loses its independence, however, loses it in the real sense that it comes under domination by men in the faculty as well as in the administration who are beholden to external powers for the power they have come to enjoy and, in some cases, for personal financial fortunes. Second, a university is supposed to educate. Yet universities are proposing that they should stop teaching undergraduates.

With the familiar irony of history, that proposal comes just at the moment when the future of the multiversity has come into doubt. You have heard cries to the effect that the recent fall-off in federal outlays has brought on a crisis in American science and in the finances of our big universities. Over the past twenty-five years the universities have received $15 billion from the federal government in support of their research programs; nearly three-fourths of those funds came from military and paramilitary agencies. I do not know who else could regret that the flood of defense and space expenditures may be receding. If this makes life hard for the universities for a while, it may also give administrations and faculties time to reflect on their proper function in our society and to make friends among the electorate whom they have overlooked in their close relations with power centers in Washington.

In plain language, the multiversity is a flop. Accordingly, we have a new Carnegie Commission on Higher Education which is reviving education as the rationale for public support of the university and has a good deal less to say about service. For the sake of the shaken universities we may hope the new sales talk will build an interest in the national budget for higher education.

But there is a deeper question here. That is the question whether any worthwhile scholarship, any worthwhile science, can be done out of the presence of undergraduates. The undergraduate compels the scholar and the scientist to argue and demonstrate the connection of his work to the total body of knowledge and to society. If the undergraduate is to understand, he must be shown what the work means and how it relates to what else we know. The scholar does not have to carry on this kind of dialogue with his graduate students; they have already bought his proposition and are his captives. There is nothing more disconcerting to the scholar or the scientist who has taken leave of his connections to the rest of us than an innocent question put to him by an inquisitive undergraduate, still seeking an education and not yet roped into preprofessional training.

The undergraduate college compels concern in a university for the unity of knowledge. Quite properly, the graduate schools are occupied with diversities of knowledge. The ongoing of our civilization requires high specialization, the digging of tunnels in all directions outward from the core of our ignorance, which is the little that we know. The further out those tunnels go on their diverging, radial courses, the further apart they reach. Yet we live in the assurance of the increasing unity of knowledge. There is only one way men know things, and that is by reason and the test of evidence. No one has charge of the secrets of nature, nor of secrets revealed by higher powers outside of nature.

Knowledge does not grow, however, by accretion of bits and

pieces, like a glacial moraine. The next new enlargement of understanding requires the reordering of the whole. Significant new questions can come, therefore, only from a firm grasp on a substantial territory of coherent knowledge. In the head of a successful innovator, that territory must reach well beyond the boundaries of his specialism.

Alfred North Whitehead said that a university does not exist to teach or to do research or to carry on the two enterprises in mere propinquity; a university exists to promote the two enterprises in interaction with each other. The aim, Whitehead declared, is to "unite the young and old in the imaginative consideration of learning."

Parenthetically, Whitehead deplored the professionalization of the doctoral degree, meaning its transformation from a recognition of capacity to a license. The course of study leading to the Ph.D., he said, is a fine way to train up mediocrity, but it is not the way to free the best to do their best. With A. Lawrence Lowell and Lawrence Henderson at Harvard he organized the Society of Fellows, a society for graduate study whose Junior Fellows are young men of such demonstrated promise that they can be turned loose to conduct their own course of study without regard to the well-worn paths to the doctoral degree.

It is out of imaginative consideration that new branches of learning—new questions in history, new insights into literature, new "-ologies" in science—find their advocates. University departments are models of the last generation's picture of how the world looked; they stand in the way of the new enterprises of the next generation of scholarship. Like the generals, the university departments are always fighting the last war.

Up to this point, I have been concerned with the implications for society of the vicious circle of incest that makes fields of learning go sterile. What the term "professionalization" implies by way of derogation is that the workers in a field of learning become preoccupied with ever more narrow intellectual concerns and with the aggrandizement of their community of interest out of any fruitful connection to the totality of human knowledge and out of relationship to the welfare of society.

Ahead of society and its interests, however, comes the welfare of the individual—in this discussion, the student—whose fulfillment is the object of society. Consider, under the heading of professionalization, what the Cox Commission on the bust at Columbia University had to say about education. The subject of education comes up just once in the report, and this is what the Commission said: "The central educational assignment of American colleges and universities has long been to prepare a functionally effective people for rather definite roles in industry, finance, government and the established professions."

If you want to understand why Columbia blew up, you need not go further than that sentence. The Cox Commission model, a fair enough statement of the university's own notion of its purpose, did not fit the self-image of the Columbia undergraduate. By now American university students have made it clear, all across the country, that quite different concerns engage their minds and hearts.

To get at what I think is on their minds I would like to enlist the help of Alexander Meiklejohn. He just missed being with us to celebrate his centenary, which comes up soon on the calendar. A president of Amherst, founder of his own college at the University of Wisconsin, and educator of a generation of educators, he lived long enough to see the American liberal arts college come and go. In a book called *Education Between Two Worlds* he had this to say:

In a government that is carried on with the consent of the governed, every citizen is both governor and governed. Men have dignity insofar as they are rulers; only insofar as they share in the attempt to advance the common welfare; and further, only as rulers have men a right to liberty. When a man is using his mind and his will in dealing with matters of public policy, that mind and will must be kept free. The public welfare requires this. That is what is intended by that magnificent first sentence of the Bill of Rights in which the government declares that not even the government shall limit the liberty of religion or speech or press or assembly or petition. When, on the other hand, a man is pursuing his own private interest when, therefore, he is one of the governed, there is no reason why his activities should be free from regulation. On the contrary, it is the very essence of government that private activities should be regulated and controlled as the public interest may require. That distinction between the citizen as ruler, who is promoting the general welfare, and the same citizen as ruled, since he is carrying on his own business, is essential to any clear understanding of what we mean by liberty in a democratic society.

In the light of that reading of the First Amendment, the university appears in its truly central place in the organization of a self-governing society. At the university the student prepares to take his place, in Meiklejohn's terms, among "the governed." In line with the prescription of the Cox Committee, he acquires the skills and the learning that society and the state urge upon him to promote the general welfare or the common defense. But the university has another, transcendent function. It is at the university that the citizen, as ruler, discovers and defines the ends of social action, the purposes to which he will one day bend the power of the government and the resources of society.

At the moral-ethical crisis of his life, the student is filled with his concerns as a sovereign awakening to his responsibility. By "liberal education" we mean an education that responds to his sovereign concern with what is right and wrong, with the promotion of justice and liberty in the land. As there is no settled way to these noble ends, there is no one program of liberal education.

The Hundred Great Books of Robert Hutchins held out the vision of a mainstream of knowledge, reaching back to the beginnings of civilization, that ought to be the common heritage of us all. When the program was installed at the University of Chicago, however, it was turned over to the second team, the college faculty that was junior in status and in every other way to the university faculty that enrolled the graduate students. Liberal education is not the transfer of a prefabricated world view, nor is it the installation of a prepared set of values.

The truly liberating function of education is to capacitate the student to know how to know. A self-governing society requires that its sovereigns be so educated that they need recognize no authority but their own judgment and no sanction but their own conscience. To make this transition to self-governing independence the student must do his learning in the presence of the primary sources; that is: the university laboratory and library and the scholars who know how to work them. To realize his independence in this company he must ultimately command a discipline.

At this point, I have talked us into paradox. Against the narrowing and isolating professionalization of learning I have argued for liberal education, which implies breadth and connection. Yet, when it comes to knowing, this calls for the discipline acquired by specialization. The greatest discovery of science is science itself. It is the disciplines of knowledge that distinguish the culture of the West from all that appeared before it in history. The paradox is one that is implied in the very name "university," in the tension stretched between the unity of knowledge and the high specialization of the disciplines by which knowledge is gained. With whatever discomfort the term implies, our universities exist to maintain this tension between collegiate and graduate education. Such considerations have led me to conclude that the truly liberating liberal education is carried on at its best in the university college, in full access to the primary sources assembled there.

Andover has resisted, for the last twenty-five or thirty years, the back pressures from the graduate schools. The school, it seems to me, must also resist the new pressure from the universities to relieve them of the task of providing liberal or college education. With all its assets, Andover does not command the necessary primary sources. There is no assurance, in the second place, that the intermediate college offerings that the school might be persuaded to undertake would be any less subject or any less susceptible to the back pressures of professionalization.

On the contrary, I hope that Andover continues to set up and maintain an offsetting forward pressure. The school should undertake all the education that can be carried on in the absence of the primary sources; it should equip its students to search out and sack the treasures of the universities to which they go. The steering committee of the Andover faculty turned in a report three years ago that said all of this more briefly and plainly. It said that the mission of the school is the development, in its students, of "the capacity to carry on higher education of the most demanding sort."

Andover is, of course, obliged also to meet its students' needs under the heading of affective education. Again, the steering committee report had the wise words:

Education is concerned with the complementary aspects of being. . . . The complement of self-understanding is the understanding of others. The complement of concern with theoretical or analytical questions is involvement with practical considerations. And the complement of personal initiative is concern for one's fellow man.

Schools dedicated to these purposes are urgently needed in American society. Andover presents a model, I hope, for schools that are to come in the country's public school system. In the first place, Andover provides for its students a much more real sample of the American community than either the suburban or the central-city high school. Our country cannot survive another generation of the class and race isolation that is promoted by the cancer of metropolitan segregation.

The suburban high school may prepare students to pass college boards and make it into their choice of colleges. But the cultural sterility of the suburbia from which they come ranks next in history only after what Karl Marx called the "idiocy of rural life." The high schools in the slums of the central cities, with police patrolling their corridors, have been little more than extensions of the city jails. That they are now graduating Black Panthers is a credit not to them but to the indomitable human spirit. A boarding school like Andover is a model for the school city, envisioned by imaginative educators, that would bring all of the young people of a metropolis together on a great university-style campus for collaboration in the drama of learning together. While engaged in such wishful thinking one might specify also that the public school cities should give learning the dignity and beauty of physical setting provided at Andover.

Andover will more fully acquit its mission of providing its students with a true model of our society as it proceeds courageously to the seeking out of diversity in the minority communities of our country and among the "low testers" deprived of opportunity. To complete the reality of the sample there is, of course, no question that this school must find its way into coordinate or co-education. Andover has a most important contribution to make to the repairing of a big hole in America's head; that is: the missing half of its intellectual power resident in the heads of women.

The discrimination that has scanted the life possibilities of half of the members of our society is so much a part of our ambience that it goes unnoticed. It begins, like charity, at home. At Harvard College, one-sixth of the applications for admission come from families with $6,000 annual income or less; at Radcliffe, one out of 36 applications comes from those homes. Andover can help raise the expectations of young women and their family's expectations of them.

Another development in American life gives schools like Andover a new importance. This is the decline of the progenitive nuclear family. If we are to have population control, we must have small families. On the other hand, the evidence is now in: the one- or two-child family has failed; it is a subcritical mass. Its failure is poignantly expressed in the hunger for community that entrains the younger generation. The weird, when not invisible, communities they create often betray infantile needs that went unrecognized in their lonely nurseries. At a boarding school students are brought together in the organization of a society for full-time engagement in the purpose of learning. From that experience they learn much else about the ethical principles of the democratic, open society that derive from the social compact. This surely is an essential experience in the education of the sovereign citizen of our self-governing republic.

. . . Now!

AT commencement on the Stony Brook campus of
the State University of New York in June, 1970, I attempted in
what follows to heal wounds inflicted by Cambodia, Kent State,
and Jackson State.

Nот everyone is attending commencement
this year. There was a time, not long ago, when no one could
afford to absent himself from this the most splendidly uphol-
stered ceremony known to our republic. I remember a vintage
year—1962—when the Secretary of Defense went back to work
in the Pentagon after a strenuous commencement season, deco-
rated with doctorates from Harvard, California, Michigan, Columbia, and George Washington. This year it is unlikely that
any member of the Cabinet, not even one so innocently employed as the Secretary of Health, Education and Welfare, will
be seen abroad outside the privileged sanctuaries afforded by
West Point, Annapolis, and Colorado Springs.

The President himself has apparently already made his closest approach to a university campus this spring. That was the 20-yard line in a football stadium before a scarcely representative sample of your contemporaries gathered by his White House chaplain.

The President chose that occasion to declare his pride that ". . . the great majority of America's young people do not approve of violence." Now, with that remark, I do not think he was telling you what you have been trying to tell him. He was saying, rather, that some of you do believe in violence; he was expressing his regret that, with the consequent suspension of civil discourse, he could not be with you today.

If I am right in this interpretation of the President's statement and if it is indeed sustained by the attitudes of any among you, then let me add my own small voice to his. The great end of liberty cannot be attained by the denial of liberty to any one. History shows, over and over again, that the means always comprehend the end. Right causes must be pursued by right methods. If that maxim has been placed in doubt by your experience—if it seems you have been getting nowhere by rational argument—then let me urge you to contemplate correctness of the maxim in reverse: violence gets violence.

Commencement, 1970, with all the new beginnings it is supposed to herald, is haunted by violence. But it took the killing of white students to shock the nation into outrage at the killing of black. Only the revelation of face-to-face atrocity by our infantrymen at My Lai has started up a proper reckoning of the vastly more numerous crimes against humanity worked so impersonally and indiscriminately against the entire Vietnamese pseudospecies by bullet, bomb, poison gas, and defoliant, delivered by helicopter gunships, Phantom fighters, and B-52 bombers. It is time, at last, to acknowledge that Indochina has uncovered no more than the tip of the iceberg. Down below, underground and under the ocean, our country—engaging the inescapable contribution and complicity of each of us—has stored up enough ready violence to terminate the attempt at civilization that began 10,000 years ago in the Fertile Crescent of Asia Minor.

If we would reverse this commitment to violence, we must

first answer the question: How can we live without it? With this question, I do not mean to imply that such preoccupation is peculiar to the United States of America. In the anarchy of nations, amplified by the present insane spasm of nationalism that has swept up the last hunters and food gatherers on earth, there is not one patriot who recoils from decoration with the fratricidal mark of Cain. My own country's foreign policy, however, is the only one I may properly—if not effectively— protest. And, for my country, I have to concede that we con- tribute more than our share to the world's sum total of ex- pended and impending violence. Over the past twenty-five years, since the end of World War II, the polity of nations has laid out $2,000 billion to hot and cold wars, to current expense and capital investment in mutual destruction. Fully half of that outlay is our country's own.

For the members of the Class of 1970 this may seem like a normally unnatural state of affairs. I am old enough to recall a time, however, when we did not maintain a giant military estab- lishment as the centerpiece of our federal government. The closest thing we had in those days to a Department of De- fense—or DOD, as its familiars call it—was a WPA, or Works Progress Administration, as I must spell out those initials to you. This is not to suggest that the solemn mission of the DOD may be likened in any way to the frivolities of the WPA. I do mean to say, however, that our lavish allowances to the DOD go beyond any rational extension of diplomacy and beyond the requirements of any war that might thinkably be fought. In the place of pyramids, some archaeologist of the next age of man may come upon a surplus missile squadron buried in the forlorn flats of what we call North Dakota and, with a turn of his spade, launch a flight of multiple independently targetable reentry vehicles upon a long-since obliterated city, half a world away. What I am saying is that the manufacture and deployment of weapons of mass destruction have assumed the same role in our attempts to regulate the business cycle that Harry Hopkins and his friends once intended for the raking of leaves, the painting of post office murals, and the hand computation of mathemati- cal tables (which, quite unexpectedly, helped to make the first atom bomb).

When I was your age federal expenditures had pump-primed the economy to bring back the top one-third of American families into the market for consumer durable goods. It was World War II that made good on Franklin Delano Roosevelt's claim that only one-third of our people were ill-fed, ill-clad, and ill-housed. And it is the cold war, maintaining federal expenditures for the DOD at a recurrent 10 percent of our gross national product, that has sustained the American celebration down to this very day. All of us, except the 25 percent or so black and white outcasts and misfits left behind in the slums of the city and the backwoods, are now redeployed on the new American frontier of suburbia. The secret to this whole historic miracle is that nobody but Uncle Sam has to buy the missiles, while the payroll that goes to the making and deployment of them arms a host of consumers, some educated up to highly sophisticated tastes, with economically effective demand for the goods and services on sale in the ordinary domestic economy.

You may think you can come up with better uses for our country's surplus capacity—for what you might otherwise call our national discretionary income. But you will be hard put to find a use with higher political expediency than wrapping it in the flag and burying it in the ground. If you, nonetheless, persist in advancing alternative uses for our affluence out of some higher and less practical motivation, be relieved to know you will find no serious objections to your idealism and general good intentions. To that same gathering in the football stadium the President rejoiced in America's willingness and ability "to clean up the air and clean up the water and provide better jobs and better opportunity and all these things for our people. . . ." You will get no argument on propositions of this kind unless you insist also upon adding a time element—like: "Now!"

That kind of mistake was made, way back in 1833, by a group of young and old Americans who assembled in Philadelphia to issue the Declaration of the Anti-Slavery Convention. They used words like "speedy," "forthwith," and "instantly." As William Lloyd Garrison said twenty years later, in 1855, "If we had spoken of slavery as an evil, a calamity, a curse to be overthrown at some indefinite period, we might have spoken in Carolina as easily as in Massachusetts; we might everywhere have been recognized as good neighbors, excellent citizens and sound Christians." Because they enunciated instead "the doctrine of immediate unconditional, everlasting emancipation" the abolitionists got no thanks when our country, after still another decade, freed the slaves at last.

If there is such a thing as a generation gap I suppose it is opened up between the young and the old by this preoccupation with "Now!" But the young have no patent on the present. It can be claimed as well, as it has been, by black Americans of all ages, who have awakened again to the discovery that they have still no other birthright but their lifetimes in our land.

Consider the persistence in American education of the elementary affront to law and morals represented by the sordid practice of racial segregation. Back in 1954, when you were just starting off to school, the U.S. Supreme Court, in the historic Brown decision, said the schools should be desegregated, but neglected to say when. A year later, in Brown II, the Court said "with all deliberate speed." In 1968, fourteen years later, Justice Brennan for the Court said that "all deliberate speed" means "now." Two years later, we can plainly see that the word "now," pronounced by the highest court of our land, did not mean then; it does not even yet mean now; for each and every one of the 3 million members, black and white, of the Class of 1970 it means too late.

There are cynical observers who find in your sense of immediacy an illustration of Dr. Johnson's apothegm on the condemned felon: "When a man knows he is to be hanged in a fortnight, it concentrates his mind wonderfully." Because your contemporaries around the world, even in countries that have no draft boards and no Vietnam, share your ethical frenzy with time, I believe you proceed from more honest premises. You have not yet got your stake in the world's prevailing order. As a lifetime arrangement, it looks less enduring than your innocence. You are right, of course. The questions of equity come foremost at this turn in history, precisely because mankind has it in its power to extend a human existence to every human being. The time to meet that obligation is Now!

VIII The Fork in the Road

The Subterfuge
of Armament

THE earliest and latest essays in this book, by date of composition, begin and end this last part (VIII). As the two essays show, I have been consistent in one view: that the commitment of our country's wealth and might to present and future wars has come not from rational perception of the national interest nor from simple aggrandizement of the military-industrial complex, but by default. In a fit of absentmindedness —or worse, in a stupor of tranquilization brought on by the spurious prosperity of the warfare economy—the self-governing electorate has failed to advance alternative claims on its surplus abundance.

Now, the revulsion of the young, the rebellion of the poor and black, disappointment in the suburbs and despair in the central cities are stirring the body politic. If the country is to find another course, conscious political decision must take over from reliance upon automatic market processes, equity must displace growth in the national goals, and the values of community and conservation must offset the centrifugal drives of rugged, devil-take-the-hindmost individualism and its reverse, the cop-out.

For all its neglect in the national theater, the political process has not failed the country completely in this past decade. The taxpayer's revolt that has been rejecting school budgets comes at the end of a period that has seen unprecedented expansion in local governmental expenditures. States, the cities, and the suburbs have accounted for the principal growth of the public sector, succeeding the New Deal, World War II, and Cold War expansions of the federal government. By such response to human need and aspiration—consider the chart of accounts in local government: education, health, welfare, police, fire, etc.— the American people worked massive redistribution of the national income, outside the economy "proper" or as the classical myth of the national identity defines it.

On different bases, in the three essays in this part, I computed the portion of the U.S. population in the public sector in 1962, in 1967, and in 1971 at one-quarter, one-third, and more than one-half, respectively. While the public sector has indeed grown over the decade, its expansion in my figures reflects differences in the major groups counted in each case. The biggest swing comes with the change in the conventions of the federal budget that abandoned the fiction of charging social security payments to an insurance fund and now acknowledges these payments as a current expense. Another major contribution comes from the tilting of the country, in effect, toward the Northeast, which has dumped the private-sector rural poverty of the South into the public-sector welfare rolls of the big cities.

Looking down the other fork in the road, no significant moderation in the arms race can be seen. The ban on testing of weapons in the atmosphere has been followed by a much higher rate of testing underground. Instead of the ten-megaton bomb conjured with in 1962, the unit of violence is the one-megaton missile warhead; 10 one-megaton weapons get more square miles of blast damage than a single ten-megaton weapon. With multiple one-megaton warheads on the United States Poseidon and Minuteman III missiles, each independently targetable, the strategy of pre-emptive strike returns to destabilize the balance of terror. Instability is compounded by the deployment of the first antiballistic missiles, unreliable and futile as these may be. The lack of progress in the SALT (Strategic Arms Limitation

Talks) negotiations presently under way suggests that the situation may have escaped management by bilateral agreement between the superpowers. Arms control and, so much more, disarmament must await multilateral negotiations including other powers, especially China.

All of this uncertainty goes to amplify the military claim on any funds freed up by reduction of the scale of military operations, if any, in Indochina. For a guide to alternative uses of such funds as might be released, the Nixon Administration could not do better than consult the special report from the Bureau of the Budget to President Eisenhower cited in the essay that follows. It was prepared under the direction of Maurice Stans, then Budget Director and now Secretary of Commerce—the secretary President Nixon forgot to introduce when he displayed his first cabinet on television.

"The Fork in the Road" was the title of the Phi Beta Kappa Oration which I contributed to the commencement of Harvard College in June, 1962. I have retitled the essay in its publication here in order to sharpen the statement of my diagnosis of the pathology in our body politic.

THE tempo of the common experience of our species is racing ahead of the biological clock. Events all out of scale with the rate and dimensions of life processes have transpired and impend.

If we had time, I would have no doubt about the outcome. But man changes and evolves more rapidly than any other invention of nature. This is because human heredity is accumulative and selective and is transmitted by teaching and learning. As the beneficiaries of this late, new phase of evolution, we cannot fail to call it by the name of progress. But, all too suddenly and unprepared, we have come to a fork in the road. The progress of which I speak has given us the means to bring every human being into full membership in the human species and shown us that we ought to do so. In the command of those same means, progress has also given the power of irrevocable decision to our historic capacity for cruelty and folly.

By one reckoning, we have just two years. There were twenty-five years, time for one generation to grow up, between 1914 and 1939. June, 1962, is not quite twenty-three years since August, 1939.

Instability is inherent in the most sensible and humane argument for stability in the present impasse in world politics. Our national security is defended, we are told, by our power to retaliate. The Soviet Union does not dare try to overwhelm us with its nuclear striking power, because it knows that we could overwhelm it in return. This is called the balance of terror. It is said to be secure against rational strategies, at least, on either side. That is to say, no statesman at present in power is likely to find a reason for attempting the first nuclear strike that would expose his own constituency to annihilation by the other side.

In recent months, however, even this insecure notion of security has been undergoing serious stress and revision. Unofficial leaks and official disclosures from the highest quarters in our government have revealed that the balance of terror has concealed a considerable imbalance. From the President himself, from the Secretary of Defense and his Undersecretary, from Senators and Congressmen and from the Pentagon—the back door, that is—we have learned that our country is equipped with a ready nuclear strike force that dwarfs the Soviet ready strike force in destructive power. In other words, there is no missile gap, nor any bomber gap, and there never was one. Throughout the eight years of their stewardship, the Republicans always stoutly denied that there was such a gap, against the claims of the Democrats. And now the Democrats are in office, and they are denying it in turn. In fact, they have released sufficient information to permit reliable estimates that our ready strike force outnumbers and outweighs that of the Soviet Union by at least five times.

To appreciate the significance of this situation requires consideration of some unpleasant technical details. The destructive power of nuclear weapons is commonly expressed on a somewhat misleading scale of tons of chemical high explosives. Thus, a one-megaton nuclear weapon is said to be equal to a million tons of TNT. Hans Bethe has calculated that a million tons is just a little less than the combined explosive power of all the old-fashioned bombs dropped on Germany in the course of World War II. To a certain extent the comparison must be discounted. A ten-megaton bomb is not 10 million times more destructive than a one-ton high-explosive blockbuster because a nuclear weapon discharges all its devastating energy at one point in space. The radius of destruction by blast increases as the cube root of increase in explosive power. You might say, therefore, that the ten-megaton bomb is the equivalent of about 200 one-ton high-explosive bombs. But this understates the case. For the destruction wrought within the five-mile radius blast range of a ten-megaton bomb is the more complete because the weapon discharges all its energy at a single instant in time.

But blast is only part of the story. The exploding nuclear bomb evolves into a gigantic fireball—three and a half miles in diameter in the case of a ten-megaton bomb. The incendiary effect of the thermal radiation from this man-made star increases as the square root of the increase in power measured on the explosive scale. In other words, the bigger bombs yield a wider radius of destruction by fire than by blast.

Thus the blast from a ten-megaton bomb will obliterate an area five miles in radius, but the heat from the fireball will incinerate an area with a radius of twelve miles. Now, if you draw these circles around Boston, with the point of your compass at the State House, you will see the central city destroyed by blast and the entire metropolitan region enveloped in fire. With the handy circular slide rule that you can purchase along with the new weapons-effects handbook, you can calculate that an attack with a total weight of about 1,000 megatons directed against the 111 largest metropolitan regions in the country could yield up to 100 million casualties. The effects of fallout may be neglected in these calculations, because the airbursts that would maximize the effects of blast and fire produce no local fallout.

My object in reciting these details to you is to show you that the civilian population is highly vulnerable to nuclear attack. This means, in turn, that a purely deterrent strike force need be of no more than modest size. Less than 1,000 megatons—a few hundred megatons—emplaced in secure and "hardened" bases have enough retaliatory killing power to keep the enemy from striking your population first. If both sides would commit themselves to a second-strike strategy, to a purely deterrent strategy, the arms race could terminate in a draw with relatively small deterrent forces on each side.

Most citizens, I suppose, have been under the impression that we have no more than a deterrent force, one that just about offsets its Soviet counterpart. It comes as a surprise therefore to realize that we are armed on a different scale entirely. Our nuclear force is of a size, in fact, that brings into the realm of feasibility another kind of strategy. The objective of this strategy is to knock out the enemy's deterrent. To appreciate what this implies, we must go back again into the technical details.

Against a hardened target, such as an underground missile-launching silo, the blast and fire of an airburst, to which the civilian populace is so vulnerable, are of little avail. The attacker must ground-burst his weapon in the hope of engulfing the target directly in the crater. When a ten-megaton bomb is employed for such a purpose, its effective radius shrinks to less than a mile. To be confident of success, an attacker must be prepared to dispatch two or more big weapons to every hardened target. The destruction of the 1,000 hardened Minuteman missile installations contemplated in our administration's present military program, for example, would require an attack with the astronomical dimension of 20,000 megatons. A hit at each target would call for pinpoint location of the target, a continent away, and fantastically accurate guidance of the missiles. The preparation of such a "counterforce" attack, therefore, implies resolute intelligence work and endless research and testing, as well as a huge preponderance of striking power.

Now, there is a school of military strategists and publicists who argue for a counterforce strategy in justification of our overwhelming nuclear superiority. From the purely engineering standpoint, they say, it is possible for us to strike first and disarm our antagonist by destroying his nuclear deterrent. We could then hold his civil population hostage under the threat of a second strike to be aimed at his cities. On moral grounds, they claim, we are entitled to such a preemptive strike because our antagonist would do it to us if he could.

But the preemptive strike, also known as "retaliation in advance," is still not a rational strategy. Its proponents concede that we would have to be ready to absorb some "acceptable" number of casualties—up to one-third of our population, say— because we cannot be sure of knocking out all the Soviet nuclear strike power. That is why the preemptive strategists are numbered among the most ardent advocates of civil defense. On the other hand, the popular apathy toward civil defense would indicate that ordinary citizens have not yet adopted this approach to the solution of the world's ills.

I do not believe that the advocates of the preemptive strike have had any significant influence on our military planning. Certainly, no responsible civilian or uniformed official of our government has ever voiced or endorsed such a proposal.

The official justification for our present military posture takes quite another line. Thanks to our superiority in nuclear strike power, it is said, the second-strike capability that would remain to us after a first strike by the enemy would be vastly greater than his first strike. But there is a contradiction buried in this line of logic that makes nonsense of this statement: an enemy so heavily outgunned could not conceivably be contemplating a first strike.

So long as the game of nuclear war is played on paper, there is never a last word. It can still be argued, it is said, that our overwhelming nuclear power promotes our security because it interdicts a first strike from the other side. But a necessary corollary to this argument is that the other side should also feel more secure in our possession of a potential first-strike capability. And in fact they have been given to understand that we would never strike first—except on some intolerable provocation.

Yet, somehow, our excess nuclear armament has failed to promote stability in world politics. The Soviet Union called off the moratorium on nuclear testing last year, and reversed the hopeful downward trend in its military expenditures. When disarmament talks resumed at Geneva this year, the Russians proved to be more than ever obsessively concerned with their geographical security and resistant to early inspection.

Our enormous armament also complicates our own approach to disarmament. We would have to do so much more disarming than the other side that ratification by the U.S. Senate begins to look like a bigger miracle than an agreement at Geneva.

Meanwhile, the prolongation of the arms race darkens the prospects of the world. If the present conference at Geneva should break down, it cannot be reconvened without the presence of China, which is on the verge of becoming a nuclear power. By that time, there will be other new nuclear powers demanding or resisting invitations to the conference. France is only the first second-class power to realize that the nuclear weapon is the ultimate equalizer, and to adopt this dangerous route back to the summit. As the number of players in the game approaches the nth number, the hazard from irrational strategies, or from mere accident, must rise. In the words of C. P. Snow, "We know, with the certainty of statistical truth, that if enough of these weapons are made—by enough different states—some of them are going to blow up!"

As citizens responsible for our self-government we are each personally confronted by grave questions concerning our responsibility in the creation of this dangerous situation. How did our nation come into the possession of such overwhelming capacity for violence? Since there is no rational military or political justification for it, we must look elsewhere for the answer.

I think the answer is not difficult to find. Beyond any doubt, the history of the last decade of our domestic life shows that the ruling compulsions were economic. The oscillations of the business cycle since the Korean War can be traced, every one of them, to variation in the rate of government expenditure for arms. Military expenditure has taken up more than half of the federal budget and fully a quarter of our manufacturing output throughout this period. After ten years of this kind of pump-priming, is it any wonder that our magnificent industrial establishment should have burdened us with such an enormous surplus of weapons?

Now we have to ask ourselves another question. How did our economic housekeeping fall into such disarray as to lay this threat to our continued existence? To approach the answer this time we must turn from economics to technology.

During the past twenty-five years our technology entered upon the era of automatic production. The real work of extracting nature's bounty from the soil and the rock, and transforming it into goods, is no longer done by human muscles, and, less and less, by human nervous systems. It is done by mechanical energy, by machines under the control of artificial nervous systems, by chemicals and by the application of such subtle arts as applied genetics. Adding up all the farmers and miners and all of the construction and factory workers, we find that not much more than one-third of our labor force is engaged in producing all of the abundance that chokes the channels of distribution. Most of the rest of us are employed in the task of distributing the abundance, keeping books on it, and repairing and servicing its working parts. To complete our census of the labor force we must face the most portentous of all the consequences of automatic production: more than 25 percent of our labor force today finds employment outside of the normal, domestic, private sector of our economy. They are employed in the arms industries or on the public payrolls, in uniform or in civilian clothes—or they find no employment at all.

Unemployment brings us back to economics. The most critical problem confronting our economic system is the insidious growth of unemployment. With each ripple in the now well-damped and administered business cycle, the number of workers left high and dry on the beach has increased. Yet, for everyone else, this has been a period of ascending prosperity. It is apparent that the disemployment of workers, by technological progress, has overtaken the growth of the economy and the now "classical" techniques for administering the cycle of recession and recovery from Washington.

Since no one can tell us how to get these surplus workers

back to work, perhaps the time has come to ask why we must find jobs for them. Surely, the aim is not to increase production! On the contrary, a little reflection shows that the objective is to increase consumption. Jobs must be found for the jobless in order to qualify them as consumers of abundance.

Our economic system and our economics are confounded by abundance because they have their roots in the history of scarcity that lies back in time behind the industrial revolution. That revolution has come suddenly to fulfillment in our life-time. We find it difficult to achieve equitable distribution of abundance precisely because our economic institutions are designed to secure the inequitable distribution of scarcity. In the more distant past, such inequity sustained the glory of civilization. Under the management of capitalism, it financed the industrial revolution.

John Maynard Keynes told the story in his famous parable of "the cake . . . that . . . was never to be consumed." Writing forty-two years ago, at the end of the First World War, Keynes observed that "the immense accumulations of fixed capital, built up during the half century before the war, could never have come about in a society where wealth was divided equitably." This "remarkable system," he said, "depended for its growth upon a double bluff or deception. On the one hand the laboring classes . . . were compelled, persuaded or cajoled . . . into accepting a situation in which they could call their own very little of the cake that they and nature and the capitalists were coöperating to produce. And on the other hand the capitalist classes were allowed to call the best part of the cake theirs . . . on the tacit understanding that they consumed very little of it in practice."

In drawing the lesson from his parable, Keynes indulged himself in a heretical forecast of the day: ". . . when there would at last be enough to go around. . . . In that day over-work, overcrowding and underfeeding would have come to an end, and men, secure in the comforts and necessities of the body, could proceed to the exercise of their nobler faculties."

For most other economists, however, and for the owners and operators of the system, the perpetual growth of the cake remained "the object of true religion." It remains so today, sustained by the almost unanimous conviction of the community that high wages are bad (because they increase current consumption) and big profits are good (because they go to increase productive capacity).

The first portent that the system had fulfilled its purpose came in the 1930's. The economics of scarcity was then confronted by the paradox of poverty in the midst of plenty. Strangely enough, as Clarence E. Ayres has pointed out, it was Keynes who saved the true religion with his "investment subterfuge." The Keynesian technique for administering the business cycle calls for increase in the current rate of investment on the downturn of the business cycle, with the government supplying the funds, by deficit financing if necessary. Investment creates consumers but no addition to the consumable surplus, and so it delivers a powerful stimulus to the entire economy. The priming of the investment pump by the government was a scandalous notion when first put into practice by the New Deal, but today it is a constitutional function of our federal government.

More investment could not go on serving, however, as the remedy for too much investment. Our economic system has found another way to certify citizens as consumers. The production of armaments, it turns out, can serve something like the same economic function as investment: it certifies additional workers with paychecks to consume the surplus, and yet it certainly makes no addition to the consumable surplus. By this device, by dumping a quarter of our industrial output into the sink of armament, we have achieved affluence if not abundance. For a few years, we even attained full employment. But now in 1962, despite a 25 percent increase in military expenditure, the number of unemployed again exceeds the number unemployed at the last recovery peak.

There are other signs that the time has come to cut the cake. The progress of technology has stirred a new ingredient into the recipe. It is the sorcerer's ingredient that so astonished the apprentice. The cake now grows out of its own substance at no cost to the abundance of its consumable output. Despite the huge appetite of the military establishment, no certified consumer goes without any good that he hankers for. Admittedly, some 50 million of us continue to be ill-fed, ill-clad, and ill-housed, but idle plant and rotting surpluses testify that we have more than enough to go around.

There is no doubt that disarmament would compel the cutting of the cake. The first word that follows disarmament in any economic essay on the prospect is: Depression. But the authors of these studies hasten on to dispel what they call "any misconceived or exaggerated apprehensions" about "the potential economic impact of an agreement." As you read on, you are enthralled to learn what promise the future holds, when we are at last disarmed and freed to cultivate the arts of peace. In the first place, both Republicans and Democrats agree that disarmament would bring no corresponding cut in the expenditures of the federal government. The major portion of the funds released by disarmament are to be invested in the enrichment of our land and our people.

In a memorandum on the economic and social consequences of disarmament addressed to the Secretary General of the United Nations, the Kennedy administration declares that this country has "a backlog of demand for public services comparable in many ways to the backlog of demand for consumer durable goods and housing and producers' plant and equipment at the end of World War II." By way of illustration, the memorandum shows that there is demand for an additional $10 to $15 billion in our annual expenditures for education, an additional $4 billion for control of environmental pollution, and $12 billion more each year for conservation and for the development of our natural resources. A parallel study by the National Planning Association sees need for a total annual investment of $66 billion in the realms of education, mass transportation, urban renewal, natural resources, and scientific research; this compares to a current annual investment of $30 billion in the public domain. To the Eisenhower administration we are indebted for a glimpse of what the federal budget might look like after a first substantial step toward disarmament. A special study by the Bureau of the Budget would reallocate $7.5 billion for education, $3.7 billion for public health, $3.2 billion for urban renewal, $4 billion for resource development, and $3 billion for space research, and a total increase in the federal civilian budget of about $30 billion. Such figures show that the Republicans can be as imaginative spenders as Democrats are reputed to be.

The Kennedy administration has yet to make such a full-dress forecast. But a report issued by the Disarmament Agency finds it possible to pick up some of the slack from disarmament by putting $9 billion into space research. With the galaxy out there beyond the solar system, we have no cause to worry about Depression!

The consensus is clear: we can offset the reduction in the arms budget by worthwhile and long-overdue investment in the upgrading of our human and material resources and the enhancement of our domestic life. The possibilities inherent in the expenditure of Pentagon-size sums on these objectives stagger the imagination.

The prospect of disarmament confronts us, therefore, with a lesson, a vision, and a question. The lesson is that the public sector—comprising the federal, state, and local governments—must continue directly and indirectly to certify a major and a growing percentage of our consumers with purchasing power. From 25 percent today, the figure is bound to go up, not down, on the day after disarmament. There is no return to normalcy in sight. On the other hand, the continued expansion in the scope and power of the government lays serious hazards to self-government. We may hope that the exercise of citizenship will commend itself in the future to a citizenry blessed with increasing leisure time.

The vision is the vision of the founding fathers: the realization of the values of freedom, equality, security, abundance, and excellence in the life of our nation.

The question is: What are we waiting for? If education should indeed command twice the present annual expenditure at some future date, then the children who are going to school today are being cheated. If our cities cry out for $100 billion worth of reconstruction in the course of a half-decade, some years hence, we are losing time and corrupting precious human resources in the slums and ghettos of the present. The same reasoning applies to the topsoil and the forests now going down the drain and up in smoke.

A hint of the answer to the question of why we are waiting is contained in the recent economic report of the Disarmament Agency which declares: "the chief obstacles . . . would be political resistance rather than deficiencies in our economic knowledge." It is difficult for anyone, including even the Secretary of Defense, to resist the demands on the public treasury laid by the armed forces. Those demands are now backed by the substantial economic interest of a giant industry exclusively devoted to armament. No such absolute moral sanction supports the claims of education, for example, and no comparable vested interest stands to gain from them. In many fields, as in natural resources and urban redevelopment, the expansion of governmental activity is bound to bring public and private interests into collision with one another. There are good grounds for the view that it will take disarmament and the threat of a Great Depression to overcome "political resistance" to our passage into the age of abundance.

But the politics of the situation can also be stated the other way: We are unlikely to get disarmament unless we are ready to embrace and vigorously advance the economic alternatives to armament. The large, round numbers I have quoted from the reports and studies made thus far must be translated into programs and engineering drawings. Local and individual initiative has an important role to play in this effort, especially in those regions and industries in which armament expenditures are now concentrated. While the federal government need not and cannot assume the entire burden, a real commitment to disarmament on the part of the administration would begin to bring the New Frontier into view on this side of the far horizon.

The choice one way or the other cannot be postponed much longer. The arms budget is losing its potency as an economic anodyne. It is concealing less and less successfully the underlying transformation of our economic system. Progress in the technology of war, as in all other branches of technology, is inexorably cutting back the payroll. With the miniaturization of violence in the step from A-bombs to H-bombs, from manned aircraft to missiles, expenditure on armaments has begun to yield a diminishing economic stimulus. Armament in any case holds out no endless frontier. By some estimates, we are already armed with the equivalent of ten tons of TNT for every man, woman, and child on earth. We acquired this monstrous capacity for destruction by a subterfuge on the investment subterfuge. There is surely little to be gained, economically or militarily, by raising that figure to twenty tons. Even in the postponement of disarmament, the economic and social consequences of abundance must soon be recognized and accommodated in our politics. If we had acquired the kind of armament most of us thought we had, scaled to the "rational" strategy of deterrence, we would be in the midst of abundance today.

In all that I have said I have dealt with the state of our nation in isolation from the world-wide political crisis that so heavily conditions our domestic existence. I have done so deliberately, in the conviction that our country's domestic situation plays no inconsiderable role in shaping the nature of the world crisis. It goes without saying that we do not command all the variables in current history. But we can and must put our own house in order or we will surely lose what command we now claim. We have come to the fork in the road.

The Stationary
State

THIS essay was the Phi Beta Kappa Oration at
the commencement of Radcliffe College in June, 1967. The
"stationary state" was a vision of John Stuart Mill; to his muse,
Harriet Taylor, we owe much of his humane influence in political economy.

In this country, today, the principal professional exponent
of the stationary state is J. K. Galbraith. In Britain it is Tory
economists who advance the idea. Labour Party economists
grant the conservationist argument against growth without end;
they fear, however, that it is not politically feasible to seek
equity without growth, for without growth it would be necessary to take from the "haves" in order to improve the lot of the
"have-nots." In the United States, at least, there is enough and
to spare in the surplus that goes to war and waste.

HALF a century ago, in his famous presi-
dential address to the Mathematical Association of England,
Alfred North Whitehead laid down the law that learning is an

act of the instant. "Whatever interest attaches to your subject-matter, must be evoked here and now," he declared; "whatever powers you are strengthening in the pupil, must be exercised here and now; whatever possibilities of mental life your teaching should impart, must be exhibited here and now."

I am lucky to be old enough to testify that Whitehead made the learning of his pupils that kind of joyful experience for its own sake. So I evoke his shade to join me here and now in the prayer that you found joy in learning surpassing even the honors you have won.

Yet I must cloud this moment with the utilitarian question. Along with the rest of your countrymen, I am wondering: What are you going to do with your Phi Beta Kappa keys? Nor does the benign shade of Whitehead steal off at the raising of this question; for Whitehead defined education as "the acquisition of the art of the utilization of knowledge." When we ask what you are going to do with your educations we ask with a concern that goes beyond our expectation of you. What roles, what careers, what opportunities does our society hold open to the educated woman? We care about you in this context because we believe in our pride that you have had the best education America affords. You present, therefore, a special test of a basic proposition of our social order: that it extends to every individual the possibility of self-fulfillment.

The America that makes such promises has been undergoing rapid—I claim: accelerating!—change throughout the years of this century. Change runs everywhere so swift and deep that it has put our country into serious confusion about its identity. The model of the American system which most citizens carry in their heads bears a weakening connection to the living reality. These considerations are relevant to our present concern because the transformation of American society has brought women into new, significant roles in its life. At the same time, the persistent and pervasive failure to come to terms with change confronts women almost everywhere with slight and obstacle to the recognition and realization of their capacities.

To all under stress of change it is promised that relief will be brought by the growth of our economy. In economics, growth is a notion as central to the thinking process as the concept of energy in physics. The U.S. economy is, of course, the model that most vividly displays what is captioned by this word. It has grown faster over a longer period of time to vaster size than any other system in the world. The national product has multiplied eight times since the turn of the century. Discounted by the nation's vigorous population growth, that still yields a threefold growth in product per capita. By dint of growth, it is claimed, new opportunities are extended to all and poverty has been all but banished from our land. The figures belie these claims, but the thrust of the principle remains.

Now physics must reckon at every turn with the second law of thermodynamics, which says energy always runs downhill. Cosmology can find no way around the prospect that the universe must come to a cold end in the ultimate dissipation and diffusion of the energy it started with. Economics is haunted by a corresponding prospect. This is the stationary state that sets in with exhaustion of the possibility of growth. The classical vision is that of Thomas Malthus: population growth would bring mankind into final equilibrium with want. During the Great Depression of the 1930's Alvin Hansen and his colleagues conjured up the stationary state out of diametrically opposite causes: the slowdown and termination of population growth would bring the growth of the American system to a halt in a permanent depression. The generation of American economists now at work has exorcised this phantom. Growth can proceed indefinitely, they claim, spurred by the genius of technology and managed in accordance with the fiscal prescriptions of John Maynard Keynes.

As long ago as 1848, however, John Stuart Mill urged his colleagues to take a more cheerful view of the stationary state. To his disposition on this issue, Mill was no doubt encouraged by Harriet Taylor, his muse and collaborator—"the friend and wife whose exalted sense of truth and right was my strongest incitement, and whose approbation was my chief reward."

The stationary state, Mill declared, "would be, on the whole, a considerable improvement on our present condition. I confess I am not charmed by the ideal of life held out by those who think that the normal state of human beings is that of struggling to get on; that the trampling, crushing, elbowing and treading on each other's heels, which form the existing type of social life, are the most desirable lot of human kind, or anything but disagreeable symptoms of one of the phases of industrial progress." As the example of this stage of civilization, in the first edition of *The Principles of Political Economy*, Mill cited the northern and middle states of America, where, he said, "the life of the whole of one sex is devoted to dollar-hunting, and of the other to breeding dollar-hunters."

Beyond the bleak contemporary scene, Mill envisioned a day when men by "prudence and frugality" might bring their numbers into rational adjustment to the earth's bounty "long before necessity compels them to it." Then, with "better distribution of property," society would exhibit these leading features: "a well-paid and affluent body of laborers; no enormous fortunes, except what were accumulated and earned during a single lifetime; but a much larger body of persons than at present, not only exempt from coarser toils, but with sufficient leisure, both physical and mental, from mechanical details, to cultivate freely the graces of life, and afford examples of them to the classes less favorably circumstanced." A stationary condition of capital and population thus implied "no stationary state of human improvement."

The U.S. economy at its present growth rate shows no sign of settling into the stationary state. Growth has brought, nonetheless, profound changes in the way we make our living and so in the structure of our society. American women, in particular, have been exchanging the preoccupation to which they were committed in Mill's time for new occupations outside the home in constantly increasing numbers. The number at work has doubled since 1940, and they now constitute more than a third of the labor force.

Growth of the national product has been accompanied by a paradoxical shrinking of the number of production workers. The muscle workers were the first to go. In 1900 they comprised a full half of all the production workers, who comprised in turn nearly 75 percent of the total labor force. In the census of 1960, it is difficult to find as many as 10 million muscle workers, and the entire production work force adds up to a 45 percent minority of the total. There are still 7 million farmers and farm hands, but 85 percent of the American output comes from fewer than one million farms. The biggest group of production workers are engaged in pushing buttons and levers and in soldering connections. In many such functions females are fully interchangeable with males. As these jobs were created, women filled them along with men. Now women are displacing men in these jobs; for the proportion of women in blue-collar employment continues to rise, while the absolute total of employment in the manufacturing industries has remained essentially constant since 1947.

The output of those industries meanwhile has doubled. Plainly, the term "production worker" must be put down as a semantic fossil. The industrial research scientist, the engineer, the systems expert, the manager, and their supporting clerical forces—all of them classified in the census as "nonproduction" employees—have made more significant contributions to the increase in output than the worker out on the factory floor. This must be unmistakably the case; for 85 percent of the nation's industrial output flows from establishments that employ less than 20 percent of the labor force, and the proportion of "nonproduction" workers in these establishments runs much higher than in the total.

It is in the economic functions still further removed from the traditional idea of production—in the providing of services—that most women have found their employment. They make up nearly half of this still growing majority of the labor force. Comparing the structure of the 1900 and the 1960 labor force, one may say that women have filled 70 percent of the jobs displaced from direct production functions by technology. Women were better prepared than men to fill the new jobs. The average American girl still has more years of schooling even though more young men go through college. Clerking and stenography became woman's work, and today women hold nearly 80 percent of the clerical jobs.

Such computations point to another novel aspect of the nation's economic growth. As the national product has increased it has become increasingly "ephemeral," in R. Buckminster Fuller's happy term. The tangible, substantial product

—for example, food and fiber from the farms—has increased. But the output of insubstantial intangibles has increased still faster, so that the overflowing abundance of American agriculture constitutes no more than 5 percent of the national product. Agriculture is now a smaller sector than either education or medical care; Americans, clearly, no longer live by bread alone.

The same trend shows up inside the industrial system in the production process itself. This has been nicely demonstrated by Anne P. Carter, of the Harvard Economic Research Project. Employing the powerful econometric technique of input-output analysis, Professor Carter set up two models of the U.S. economy, one embodying its technology as of 1947 and the other the technology of 1958. The two model systems, installed in a computer, were set the task of delivering the same 1958 gross national product. Comparison of the inputs required by each produced a quantitative measure of the increasing ephemeralization of the production process. The 1958 system, on the one hand, absorbed 32 percent less materials and, on the other hand, required a 12 percent larger input of such intangibles as energy, technical and financial services, and communications. These "general inputs" are essential to the production, sale, delivery, and servicing of the tangible product, but they are not incorporated in it. In the 1958 technology such inputs represented 55 percent of the total requirements of the industrial system.

That is a somewhat roundabout, but precise, way to say that the work of people in the U.S. economy these days involves them less and less with things and more and more with other people. Moreover, as the changing make-up of the labor force indicates, the process of ephemeralization appears to favor the employment of women.

Another development that favors the employment of women is the expansion of the public sector. This is a development that troubles the nation's self-image. In 1900 no more than 6 percent of the labor force were employed in government. The nation then still approximated its identity as one engaged in the pursuit of happiness; that is: the pursuit of private interest, the citizen's own or his employer's. Today 14 percent are employed directly by the federal, state, and local governments. As these payrolls have increased over the years, the proportion of women employed by them has increased. While women constitute 36 percent of the labor force as a whole, they hold 43 percent of the jobs in the public domain. The most important increases in the public payroll and in the employment of women on it have come in recent years from the growth of state and local public agencies.

All told, 11 million people are directly employed by government in the United States. But this does not exhaust the employment generated by public expenditure. In the first place, the accounting thus far has considered civilian employees only. The public payroll employs—if that term can be applied to national service—the 3.4 million in uniform. A full accounting must consider also the indirect employment generated by the expenditures of public agencies in the private sector. They lay out about half of all their funds for procurement of goods and services. They thus provide jobs for as many more people again as the civilian payroll of the government. Expenditures for public purposes, altogether, employ a full third of the labor force.

Of course, expenditures for military purposes now account for nearly one-third of this one-third. But the experience of the Great Depression, stated explicitly in the Full Employment Act of 1946, charges the government to create jobs where the market operations of the private economy fail to do so. This is a declaration of public policy that cuts across the American tradition of minimal government.

There is a suspicion, deep in the political subconscious, that economic considerations as much as military have maintained the national defense budget at around 10 percent of the increasing gross national product ever since 1950. It has proved easier to invoke the absolute of national survival in order to rationalize giant public outlays than to sustain the collisions with private interests that have confronted expenditures for a New Deal, a Fair Deal, a New Frontier, and a Great Society over the past thirty-five years. In this interim, however, the peculiar institutions created to manage military outlays—the great aero-

space companies, the "nonprofits," and other hybrid enterprises—have been obliterating the boundaries once so sharply
drawn between the public and the private sectors. These precedents stand ready to help mediate any prospective large-scale
shift of public outlays from military to domestic purposes.

All the evidence thus compels the conclusion that the market
—that is, the traditional model of our society—holds a waning
influence in the nation's life. No one can accept this turn of
events without anxiety and regret. The play of adversary forces
in the market has provided the only universally sanctioned
standard for the objective determination of value. What is
more, the making of innumerable private choices, when added
up, could be taken as the assertion of the public interest.

It is true that the market has not realized the equities
promised in the equilibrium of the classical system. The real
world, Joseph A. Schumpeter once remarked, is the outcome of
short-run processes that never make the long run to equilibrium. The distribution of incomes and assets in the U.S. population has not changed significantly, despite the radical changes
in the way Americans make their livings. Growth, not increase
in equity, has reduced poverty to an affliction of a decreasing
minority of the population. Hope has sustained faith. And faith
in the ultimate equity of the market transcends economics.
Americans equate their personal liberty with the freedom of the
market. In Justice Holmes's words: "The best test of truth is
the power of the thought to get itself accepted in the competition of the market."

Yet the country is now deploying one-third or more of its
human and material resources outside the market, in accordance with public rather than private choice. The people have
reason to pay renewed concern and attention to the task of self-
government. By what objective standards can the public interest
be determined? What procedures can win for public decisions
the confidence accorded to the market process? The military
agencies have developed highly disciplined and powerful techniques for the planning and accounting of large enterprises.
"Benefit-cost analysis" and "program planning and budgeting"
have drawn the admiration of administrators in the socialist

countries. But these remain techniques for evaluating, not making or yet framing choices. The celebrated interplay of countervailing forces fails equally to provide a surrogate for the market process. In the classical market, the consequences of individual choices are damped out in the drift of the larger system. The public interest has no such inertial insulation in the contention of forces marshaled by Big Government, Big Labor, Big Business, and, lately, the Big Military.

This is the moment to recall Alexander Meiklejohn's dissent from the dictum of Justice Holmes. Meiklejohn placed the right to be able to listen ahead of the right to speak. "We Americans," he said,

> have taken the "competition of the market" principle to mean that, as separate thinkers, we have no obligation to test our thinking to make sure that it is worthy of a citizen who is one of the rulers of the nation. . . . The public interest cannot be merely the totality of private interests. It is, of necessity, an organization of them, a selection and arrangement, based upon judgment of relative values and mutual implications. . . . The meaning of our common agreement [is] that, working together as a body politic, we will be our own rulers.

Such reflection is relevant to our consideration of the role of educated women in American society. They are finding their careers largely in the professions and in the expanding public sector. Generally speaking, as earners of single-person incomes or of second incomes, women are more free to follow a profession as a self-rewarding vocation. Against the gradient of economic pressures, they turn up, time and again, in unexpected and neglected offices where they make original contributions to their professions and to society.

Women are represented in the professions in somewhat smaller proportion than in the labor force as a whole. The "professional and technical" is the fastest-growing category of employment; it is now 12 percent of the total, and the number of men counted in it has been increasing faster than women in recent years. In the learned professions men greatly outnumber women, and their median incomes run two and three times higher. Women tend to disengage from the entrepreneurial struggle of self-employment. Seeking salaried positions, they find them in the government and in nonprofit institutions. There, many women serve crucial functions that have been otherwise overlooked in the allocation of economic incentives.

Of the few women who are physicians (only 6 percent of this profession) a significant percentage hold immensely responsible and grossly underpaid jobs in local public health agencies. Others hold staff appointments in hospitals, in school systems, and in welfare centers, where their services are urgently needed and equally underpaid. Similarly, the even fewer women lawyers (3 percent of the profession) hold salaried positions in government, in local social welfare agencies, and in legal aid or public defender offices. Presently, as problems of urban life, poverty, civil rights, crime, narcotics addiction, social welfare, and the delivery of health services come to the center of political attention, it is being realized that these women have been there all the time. They are providing leadership and a resource of professional understanding for the development of new public policy.

Teaching, the biggest of all professional occupations, is the only one in which women predominate. This does not apply, of course, to teaching in universities and colleges. If we might identify the tenured professorships in our hundred greatest universities as a "profession," then this is the profession that slights women most meanly; my own cursory research, starting with the Harvard catalog, shows women fill not more than 1 percent of these positions. In elementary and secondary schools, however, 1.4 million women teachers fill 70 percent of all the jobs. Teaching at these levels has been regarded as woman's work and paid accordingly. If values were properly distributed, the pay scale in education might be turned upside down to provide a full professor's salary to the nursery school teacher who knows how to get her pupils off to a running start. In fact, both the status and the salary of the schoolteacher have been leveling upward as the expansion of the country's school systems has generated the principal recent increases in employment in the public sector and in the economy as a whole. The teachers are welcoming the men now attracted into the school systems and have no regret at the desegregation of their profession. They always knew their work was important, and they are gratified to have others join them in that opinion.

The material abundance of our economy that engages ever fewer workers in its production sets ever larger numbers free to realize the possibilities thus secured for human life. At present, the stationary state remains a far-off prospect. Yet millions of Americans are able to turn from coarser toils to occupations that exercise their higher faculties. The growing ranks of technical and professional people are finding increasing employment on the public side of the fading boundary between the public and the private sectors. Education is now on its way to becoming the hugest enterprise of our society in terms of invested capital, expenditure, and number of people. It holds the best promise that Americans will organize themselves in a body politic and be their own rulers. Women have been engaged in this civilizing activity in greater numbers and for a longer time than men. The nation is getting ready to acknowledge that two heads are better than one.

Time to
Cut the Cake

THE pseudoscience of futurology anchors its predictions in misplaced concreteness. With the numbers to plot a curve out of the past it finds the future by extrapolation; thus, we are admonished in a study sponsored by a distinguished foundation to make ready for 147 million automobiles, for electrical-power output that will require 10 percent of the continental surface runoff for cooling water, and so on—all by the year 2000. No law of nature or society requires, however, that statistical curves maintain their curvature. No trend can stand on end indefinitely, least of all a trend plotted by the exponential curve that is celebrated in the first essay in this book.

In the gothic spirit of the soaring curve, Americans have relied upon growth to heal their differences. Though inequity persists in the distribution of goods and opportunity, growth has reduced the numbers of the acknowledged poor to 30 million. Against the mutual congratulations all around that celebrate this "progress," it should be asked how the richest industrial nation can permit the growth of some numbers of children in its population to be irrevocably stunted by malnutrition. There are nations not as rich that do not. Poverty, moreover, is relative—a state of "low-income" as the U.S. Bureau of the Census now calls it. The number of Americans who feel excluded from the celebration, however, goes far beyond the official 30 million. The nation is riven by differences that growth will not heal.

At Washington University in November, 1970, on the Ferguson lectureship "on some aspect of science," and at the Santa Cruz campus of the University of California in February, 1971, on a Regents lectureship, I assembled the elements of the essay that follows. I reach here for an alternative to growth, especially growth sustained by overkill.

A bench-mark test of our country's progress toward the alternative of equity is presented by the priority accorded to the guaranteed annual income (see "The Sorcerer's Apprentice," p. 124) on the political agenda. The cheerful view I set out in what follows has been dampened by subsequent events.

One event was Wyman, Commissioner of New York Department of Social Services, et al. v. James, the maiden decision of Mr. Justice Blackmun for a six-to-three court. The ethic of the nineteenth-century parish workhouse board supplies cold, hard lines to the new justice's prose. Rejecting Mrs. James's claim that she need not submit, as a welfare recipient, to the indignity of "home visitation" by the welfare department, he writes: "One who dispenses purely private charity naturally has an interest in and expects to know how his charitable funds are utilized and put to work. The public, when it is the provider, rightly expects the same. It might well expect more, because of the trust aspect of public funds, and the recipient, as well as the caseworker, has not only an interest but an obligation."

And again: "We note, too, that the visitation in itself is not forced or compelled, and that the beneficiary's denial of permission is not a criminal act. If consent to the visitation is withheld, no visitation takes place. The aid then never begins or merely ceases, as the case may be. . . ."

In his dissent, Mr. Justice Douglas declared ". . . the central question is whether the government by force of its largesse has the power to 'buy up' rights guaranteed by the Constitution." To which he added: "If the welfare recipient was not Barbara James but a prominent, affluent cotton or wheat farmer receiving benefit payments for not growing crops, would not the approach be different?"

The other dampening event was the distortion beyond recognition of what had amounted to the guarantee of an annual income that had found its way into the welfare reform bill of the Nixon administration. As this reform comes from Congress it is giving full legislative embodiment to all the Dickensian ghosts that haunt the Blackmun opinion in Wyman.

THE American economy presents the disorderly spectacle of concurrent depression and inflation. Unemployment has not stopped rising and, at the same time, the cost of living goes on increasing. The system has not been responding to the controls. After two years of tinkering by monetarists, the administration has gone back to the old "New Economics." It is still an open question whether the expansionary budget floated in this year's State of the Union message will kick off a new surge of growth or just more inflation. People in every income group are seized with insecurity; our society is divided by fear and upwelling ugly prejudice; political tensions are stretched between the poles of left and right.

To find the cure for the land's malaise, it is apparent, we must look outside the terms of reference of economics. My diagnosis is easily stated: Progress in science and technology has outrun the pace of evolution in our social institutions. I hope to make this a more useful statement by examining the evidence at close range. The detail will be sufficient to demonstrate, for example, one brake on the evolution of society; this is the stake that even wise and generous men may acquire in the arrangements prevailing at the time they arrive on the scene.

The application of new scientific knowledge in technology makes production in industry after industry increasingly automatic. *Disemployment* must be reckoned with, therefore, alongside unemployment. While the current history of the economy proceeds in cycles, those cycles can no longer be regarded as merely cyclic. They reflect underlying progressive changes in the way Americans get their livings. The system has shown considerable resilience in the redistribution of the dis-employed to new kinds of employment. It must be recognized, however, that most new jobs are now being created outside the economic system proper, in the public sector. Deep changes are running in the character of American society that are not yet recognized in the myth of our national identity.

From this diagnosis there follows the prescription here sum-marized: This country can no longer seek relief of its domestic anguish by the pursuit of economic growth; it must face at last the issues of equity that divide the people. In the place of jobs in its economic calculus it must establish individual fulfillment —the self-realization of each citizen in his role as a contributor to the wealth and life of society as well as in his role as con-sumer and beneficiary. Against private interests it must assert the claims of the community. Finally, it follows, our society can no longer rely upon the market to objectify value. Much of the country's total economic activity now proceeds outside the market. Many market transactions transcend the interests of buyer and seller. Our democracy must learn to make rational and witting choices by political processes scaled to the dimen-sions and consequences of each choice.

It is possible to measure quite quantitatively the widening gap between science and society, for it may be defined as the gap between the productive capacity of our technology and the capacity of our institutions to find uses for it. The yardstick is the difference between what we produce and what we consume. In the classic economic model, this difference is put away in savings by individuals to furnish the capital for investment in economic growth. History shows, however, that most of the saving, for whatever purpose societies have had for it, has been involuntary, compelled by institutions designed to secure the inequitable distribution of scarce supplies. Automatic produc-tion requires redefinition of this excess of production over consumption. In the full measure of its abundance, it may now be regarded as our discretionary income. The telltale question is: What do we do with it?

The figures show that we customarily set aside 20 percent of the gross national product (GNP) in public and private investment. Except for the bite of federal taxes, it can be said that people scarcely notice this subtraction from their current consumption.

To the investment figure we ought to add the military budget; for the product of this outlay does not find its way— apart from such occasions as Kent State—into current domestic consumption. It serves, however, as a surrogate for investment and has maintained employment and consumption in the consumer economy when investment has faltered. Our discretionary income, thus defined, comes to 30 percent of the GNP.

If we accept the puritanical definition of production employed in the national bookkeeping of the socialist countries, the U.S. discretionary income comes out still larger. We must reckon-in all of the domestic welfare expenditures of our federal, state, and local governments, plus private expenditures for such purposes as entertainment, education, and medicine. Significantly, these have been the fastest growing sectors of employment in both the private and the public payrolls. The discretionary income now takes up 52 percent of the GNP.

A still more strict accounting would add to this 52 percent the waste that so characterizes the affluence of America. A substantial portion of the waste is accounted in the outlays for trade, distribution, and services and is visible in the cost of sales, even granting that goods must be pushed in an economy approaching surfeit. To the visible waste must be added also some portion of the expenditure for packaging that bulks so large in our annual output of 3.5 billion tons of solid refuse. Less visible and harder to account are the tributes paid to phoney "R & D," to useless model changes, to planned obsolescence, to extravagant use of materials and energy in the design and operation of the consumer durable goods that adorn our culture. With these additions conservatively estimated, the discretionary income may be rounded upward to 65 percent of the GNP.

One rough test of the validity of this reckoning is provided by the occupational distribution of the labor force. The production sectors of the economy—agriculture, mining, manufacturing, transportation, and communication—employ no more than 40 percent of the labor force as production workers. No matter how you might redefine the somewhat archaic definition of "production worker" employed by the Department of Commerce, you still have 60 to 65 percent of the labor force available for other, non-production and, so, more or less discretionary activities.

From the total of our discretionary income, this accounting shows, our society lavishes 23 percent on war and waste. For too many of our fellow citizens, this outlay for war and waste produces the byproduct: want.

According to the American plan, economic growth was to eliminate poverty. Growth of the product would ultimately employ everyone and so qualify all as consumers of the goods they would produce in abundance. This has been the faith not only of the Chamber of Commerce boosters who, in Adam Smith's phrase, "affected to trade for the public good"; it is there also at the core of the neo-Keynesian prescription that valves in investment to level out the business cycle. Thus, in a cadenza composed to acknowledge the bestowal of his Nobel Prize, Paul Samuelson wrote: "The art of economic policy is that of proper priorities and quantitative dosages. . . . The highest priority should now be given to restoring vigorous real growth." John Maynard Keynes had no such commitment to growth-without-limit; he could envision the day when "the cake" might be cut.

The growth economists have failed consistently to reckon with technology. Despite all attempts to read the numbers to the contrary, progress in technology does eliminate jobs. In the past, the private sector has picked up the slack by inventing jobs in non-production functions. It was here that private sponsorship of waste played its economic role. Over the last 25 years, the burden of job-invention has fallen increasingly to the public sector. More than two-thirds of the 10 million new jobs created between 1960 and 1970 were generated by the expansion of governmental expenditures, half of the jobs directly on the public (especially the state and local) payroll, and half the jobs indirectly on private payrolls created by government purchases. It is here that public sponsorship of war plays its decisive economic role.

The pay check, however, now certifies a declining percentage of the population as consumers of the abundance. Massive income transfers by social security and welfare checks qualify more than 30 million payees with purchasing power; that is more than twice the number of people who earn their livings on the public payroll. Taking these 45 million public-sector people, adding those employed by public purchases in the private sector, and counting the dependents of all of these, we come to an historic number. It is a watershed number of more than 100 million; it says that more than half of the American people now depend for their daily bread on money laid out for public purposes.

Our economy has attained enormous growth. Going beyond the premises of the American way, it has pushed the redistribution of incomes through channels outside the private sector. Yet the poor are still with us. The public sector remains the poorhouse of America. That is why it is able to support more than half the population on no more than 35 percent of the GNP. Growth has not compensated for inequity, nor has the redistribution of income reduced the old inequities in the allocation of shares in our economy.

The one significant change in recent years (since 1952), according to Joseph A. Pechman of the Brookings Institution, is a regressive one: a gain from 30 to 34 percent in the share going to the top 15 percent income group. This is, by definition, a gain at the expense of the 85 percent below. It is of special interest because it went in its entirety to the 14 percent below the top 1 percent (whose 7-percent share remained untouched). These percentiles are identified by Pechman as consisting of "most of the professional people in this country (doctors, lawyers, engineers, accountants, college professors, etc.), as well as the highest paid members of the skilled labor force and white collar workers." For the rest, the classic inequities obtain, with the bottom 20 percent getting by on less than 5 percent of the national income. Some substantial portion of that bottom 20 percent, an officially acknowledged half of them at least, are sunk below the poverty line in the defensive culture of poverty that is mistakenly identified with race.

The application for admission to the wider society coming

358 The Acceleration of History

from rising numbers of the poor and black sends shock waves of anxiety through the next five 10-percentile layers just above, who share with them the bottom half of the national income. To these middle Americans, fellow-citizens less fortunate than themselves now appear not as objects of concern but as the example and threat of deprivation and humiliation yet to come. The decal flags on the windows of their cars signal the danger of the gorge rising in tens of millions of throats.

As we apportion concern among our fellow-citizens we should keep some in reserve for those in affluence at the top of the income pyramid. They are discovering, perhaps too late, how little they get for so much. Our rubber-tired transportation system has carried into the third quarter of the twentieth century the social aims and the personal morality of *laissez-faire*. The pursuit by each of his own has brought all into collision with one another in what Keynes called the paradox of the aggregate. What was so desirable to each proves upon attainment—in the traffic jam on the freeway, in the crowded vacancy of suburbia, in the tedious lunacy of the TV screen—to be frustration for all. Each has bought in "his own television set and private automobile," says E. J. Mishan of University College, London, "the elegant instruments of his estrangement from others." There are no friends to turn to when, as now, a recession betrays the insecurity of high-salaried income from services of uncertain value rendered to impersonal corporations and governmental agencies.

The tragic Lyndon B. Johnson, in his ambition to go down as a great domestic President, sought to make equity a working political issue. His first thirteen months in office, filling out the unexpired term of John F. Kennedy, carried to its high-water mark the thrust of the welfare state established in Washington during the 1930's. In those few months, Johnson ran through Congress the entire unfinished agenda of the New Deal. The best that his "in and out-er" university consultants and civil-service counselors could come up with thereafter was the tax cut of 1964. This finger exercise in fiscal management, advertised as the formal installation of the New Economics in the management of the national household, was supposed to lift the economy from the slide that had been depressing the outlook for Kennedy's re-election. Compared to an equivalent unbalancing of the budget by expansion in welfare enterprises, it was a sufficiently dubious measure. What picked the system up and carried it onward to its latest burst of growth was the commitment of the nation's power to full-scale war in Indochina.

The venture into Indochina cannot be comprehended without reference to its significance in the management of the economy at home. It is another turn in the epicycle of military spending that has driven the business cycle throughout the twenty-five years since the end of World War II. The Korean War provided the necessary external stimulus at the moment when the system approached exhaustion of private domestic demand that had been pent up during World War II. During the sunny Eisenhower years, the start-up of the capital-intensive strategic weapons systems—"More Bang for the Buck"—sustained the military budget at an average 10 percent of the GNP and spared the Republican party the need to call up the labor-intensive infantry. That political embarrassment has been reserved in this century for Democrats.

By the time John F. Kennedy summoned his fellow-citizens to place one of their number on the moon, the economic impact of the revolution in military technology was waning. It is during their start-up that these great systems undertakings deliver their principal stimulus. The space program had already imparted its giant step to the business cycle, therefore, when Neil Armstrong took his small step from the bottom rung of the ladder. Vietnam had meanwhile brought a million-man increase in uniformed employment and a corresponding increase of employment in the war-supporting industries.

No exigency of world politics comes near to justifying the chronic commitment of 10 percent of our GNP to war and preparation for war. We have succeeded in frightening our friends as much as any of our putative foes. If we are really willing to understand why we make such terrifying misuse of our discretionary income, then we must face some unpleasant truths about ourselves. It is not only that the military-industrial complex wields such influence in every Congressional district; it is not only that the surplus finds its way more easily into waste than into use; but it must also be admitted that no substantial element in the electorate has effectively advanced any other public claim upon the surplus.

Like the good Germans who lived upwind from Dachau, most of us have our interest, willy-nilly, direct or indirect, in the war and the cold war. This method of keeping America going has had something in it for almost everyone. Until Vietnam brought home the truth that war expenditures are for the making of war, the moral conscience of the nation was tranquilized by well-being. What is more, no one enjoyed a bigger increase in well-being than those elements in society who are generally regarded as the repositories of its conscience.

The deployment of the military budget contributed decisively to the improvement in the incomes of the 14-percentiles just below the top. That is the socioeconomic niche of the "scientific-technological elite" whose prospective "undue influence" excited the suspicion of President Eisenhower. In the universities and in industry, as well as in the government proper, military expenditures have employed or financed the work of two-thirds of the country's scientists and engineers. Professors of the social and political sciences also found their way into the fortunate 14 percent during this period; they did so as contributors to the extensive literature that divides mankind into two worlds of the free and the red, sets the strategy for counterinsurgency operations and prescribes priorities for post-attack recovery.

In a simplistic way, perhaps, too much can be made out of the connection between military expenditures and the exertions of these scholars. On the other hand, if black studies, urban affairs, technology assessment, and like topics are now listed among the preoccupations of the social and political science departments, this is by the action also of equally external initiative and funding.

It is high time that our universities got free from their now thirty-year-long enchantment with Washington. Owen Lattimore, who has observed the process from a special vantage point, set down a perceptive analysis of the resulting subversion of the universities and their scholars:

During the war the professors, and the students who would one day be professors, had been drawn into government and military service in numbers unprecedented in American experience. Many of them were flattered by the way their techniques of research and analysis were meshed in the gathering of intelligence and the drafting of policy. They were gratified when, after the war, they were called back to Washington for consultant work and shown "classified" information, under the insidious formula: "Read this, but don't show it to anybody. Let it guide what you publish, but don't tell anybody why your work has begun to take this direction, this emphasis."

. . . [T]hat explains why the government contract is one of the most corrupting influences in American academic life at the present time. Under this system the government supplies a university with money and "classified" information. The pseudo-"private" scholars submit the results of their work to the government. Because of this confidential relationship either the work as a whole is barred from the public market of knowledge, ideas and criticism, or only part of it is released to the public—which is just as bad. This way of doing things inflates the arrogance of bureaucrats, by making them more immune to criticism, and increases the timidity of scholars, who hesitate to talk or write freely about their work, for fear of revealing their sources of information and being blacklisted as "security risks."

The input of disinterested scholars and thinkers is needed if the next turn in our affairs is to be one for the better. For it is clear that the delegation of decision to the absolute of national security and the play of self-interest has failed. Dissent, protest, discontent, and disorder; entraining draftees and veterans; students and nonstudents; blacks, Chicanos, Amerindians and ethnics; rentiers who find their beaches (as at Santa Barbara) polluted by the springs of their own dividends and wage earners who see their take-home pay eroded by inflation; doctors and patients; duck hunters and bird-walkers; welfare recipients, luxury-apartment dwellers, and suburban commuters; consumers, policemen, grape pickers, hardhats, airline-ticket clerks, homosexuals, women, and men—all testify that the nation is ready to find a better way to manage its household.

This means the choice of more constructive uses for our discretionary income. The application of any portion of it to the redress of inequity can do more to reduce prevailing political and social tensions than a much larger growth in GNP.

The most plainly isolated and readily identified portion of the discretionary income is that at the command of the Pentagon. No one has suggested that the entire military budget should be rededicated from the common defense to the general welfare. Congressional investigations have shown, however, that efficiency could save at least 10 percent of it. More rational calculus of the balance of terror could save another 20 percent. An arms control treaty could reduce the burden by a full 50 percent.

The last serious work on the redeployment of military dollars dates back to the early 1960's. At that time, studies by the Bureau of the Budget, the National Planning Association, and a group of university economists sponsored by the Friends Service Committee showed that the neglected public home front could absorb all or virtually all of a 50 percent reduction in the arms budget, with little of it available for any tax cut.

There are many audible and visible takers for the $40 billion that would be freed up at the present rate of military expenditure. John V. Lindsay has said that New York City could use $5 billion a year in additional capital investment for an indefinite time to come; the nation's new housing construction is running short at an annual rate of $10 billion; such a one-time undertaking as the depollution of the Potomac, a river of modest length, flow, and density of polluting population, has been estimated at $7 billion. A Carnegie commission on higher education has calculated that higher education in America—meaning the eminent privately endowed institutions of the Northeast as well as the great state universities of the West—will require a federal subsidy of $13 billion by 1975. To enhance the services, resources, and amenities of the public sector in the inner city and in the national wilderness offers the straightest route to equity. The public sector—the commons—belongs to everyone.

Wassily Leontief, with the precision of input-output analysis, has shown that a military budget cut of $1 billion must be offset by a domestic expenditure of $1.15 billion if employment is to be maintained at the same level. This is because it is more expensive to keep a young American male at home in the style to which he is accustomed than in a barracks in New Jersey or a foxhole in Vietnam. The growth economists do not, therefore, have to yield the field all at once to the new and not yet systematized economics of equity.

It is easy enough to write up a shopping list for the spending of the funds that may be pried loose from the military budget. Something more is required to secure a corresponding public benefit from such expenditures. Housing would come high on almost everyone's list, urban housing to be made available at a rent people can afford. On Manhattan island the cost of construction, inflated by wage scales that reflect the easygoing economics of conspicuous consumption in the erection of corporate headquarters, has priced new dwellings for people beyond the reach of all but the top percentiles of the upper 14 percent. Subsidies and other offsets to such costs necessarily bring public housing into conflict with private interests or under their control. The last major effort on this line, under Title I of the Housing Act of 1949, yielded a net reduction in the number of urban dwelling units and put the new dwellings subsidized by public money on the market at even higher rents than those gouged from dwellers in the slums that were cleared. A review of this history shows why it is politically so much more expedient to turn our discretionary income over to the military; they remove it from domestic politics by interring it or bestowing it overseas.

A similar lesson is taught by the current experience with Medicare and Medicaid. Federal subsidy of health services for the medically indigent elderly and for recipients of public assistance and their children induced an instant inflation, now exceeding 50 percent, in the costs of medical care. The result was to swamp the third-party payers, raise the level of medical indigency, and narrow the door to preventive and early care.

Plainly, a change in the moral climate of our society must attend capture of a larger portion of our discretionary income for effective redress of prevailing inequities. Such change already drives some of the most novel crosscurrents in American life and politics. The consumer movement, identified with

Ralph Nader, reflects widespread disillusionment with the ways in which the private sector disposes of its portion of the surplus. Even the legions of the worried and frightened who would shut off dissent on other questions are ready to support measures that will curb waste and fraud by manufacturers and their distributors. With competitive imports available, Americans are showing increasing resistance to the sale of fantasy and status instead of performance in consumer durable goods. A perhaps more altruistic sense of the public interest is expressed by the very large number of citizens who have learned the word ecology during the past few years. Their declared willingness to absorb the cost has encouraged manufacturers to internalize the diseconomies of polluted air and water.

The inner cities, it is encouraging to note, provide the setting for the most explicit demonstration of a rising sense of community among us. Community action groups confront the inertia of vested interest in the public as well as the private sector. In New York City they have compelled the decentralization of the public education system, placing the schools under local boards more accessible to the parents and others who share their concern for the children. Community health groups are similarly persuading the medical schools and voluntary and public hospitals to make their services more readily available in circumstances more regardful of the dignity of the individual.

Hope for resolution of the paradox of the aggregate hangs upon our recognition and response to the imperatives of the increasing mutual interdependence of individuals in the communities of industrial society. The attempted escape to suburbia has only extended the reach of our essentially and necessarily urban communities. Whether community and private interest can find happy resolution in the American soul will probably be tested first in the realm of transportation. Urban rapid transit comes at least second on every shopping list prepared for the redeployment of our discretionary income. There are signs that increasing numbers of auto-driving commuters will exchange personal convenience on the first and last laps of their journey for less wear and tear on their nervous systems in the traffic jams between. The political contest impending here is substantial, as is indicated by the crippling of the biggest present venture in rapid transit, the BART system that is to connect San Francisco to its suburbs. By default of our sense of community, we have created a stubborn political interest in the construction of highways that presents each year a giant capital subsidy to the rubber-tired transportation industry.

A corresponding transformation is required and is even under way in our moral attitude toward work. Nothing so demonstrates the persistence of the Puritan ethic as the still-prevailing notion that a day's work must be a disutility endured in exchange for the utility of a pay check and the things it buys. The reciprocal relation between work and pay is best enforced, in the popular view, if the work is performed for the profit of another. That other one's interest in his profit is thought to insure that he will exact of his employee the full disutility of the day's work. Since public employment places the employee in direct adversary relationship to the taxpayer—even if he be one and the same person—it is not surprising that the gravest suspicion attaches to public employment and that low morale prevails on the public payroll. By so much more do the suspicion and contempt of the community fall upon the recipient of the welfare check, who must render up the disutility of shame in return.

Change in the moral climate that surrounds getting and spending is evident, however, and is being forced by circumstances. It is no longer pretended that social security checks are paid from an insurance trust fund. In the government's accounts, the payments are now stated as simple expenditures from the federal budget; they are barely balanced by "employment taxes." It is acknowledged that the recipients of social security have every right to their incomes, and it is universally agreed that those incomes are inadequate.

Something like the same pragmatic attitude is beginning to develop around welfare payments. Milton Friedman of the University of Chicago, out of his old-fashioned regard for the dignity of the individual, wants the welfare system abolished in favor of a negative income tax—that is, payments to the individual on a sliding scale inverse to earned income up to the point where the sign reverses and his earned income begins to yield tax payments to the government. In Friedman's world, the poor would be free to articulate their needs in effective economic demand and to live in Gin Lane, if they pleased, without the paternalistic interference of the welfare worker.

The administration's welfare reforms go even further toward breaking the spell. They propose the payment of a minimum income to every family lacking other means of support. The largesse is hedged and conditioned, however, by provision of "self-help" incentives and compulsions upon the beneficiaries to seek gainful employment in jobs that may or may not be there.

What makes halfway measures fail is the moral taint of compulsion that hangs over from the bad old days of scarcity. Then, it took the goad of necessity to get the community fed and the knout to gather in the harvest. Today, automatic production has made these compulsions obsolete. They ought to be stricken from our public laws and private ethics. A modern industrial society can guarantee a humane and decent existence to every one of its members. It should extend such guarantee to each as a matter of right.

Having disjoined incomes from compulsion our society should do the same for work. While no contemporary society gets its work done without economic incentives, the American market does not validate with incentives all the different kinds of work people want to do and that the community needs to have done. The disjunction of work from compulsion, freeing people to do things that are worth doing for their own sake, could bring on a renaissance of crafts and a liberation of creative enterprise into unexpected new realms of the arts. The counterculture might soon fire up a new countereconomy. But the avant-garde must maintain a long lead and Off-Broadway must go Off-Off if its practitioners are to escape co-optation by the surrounding and all-embracing system.

It is in the neglected commons that work most wants doing. The kind of action needed to get action going here has been dramatized by strikes-in-reverse—after the example of Danilo Dolci's people in Sicily—conducted by the Young Lords, Black Panthers, and Blackstone Rangers. They have set up crèches for the infants of working mothers, have conducted community health surveys and propelled active tuberculars from their tenements into proper medical hands, and have organized community backyard clean-ups that have exposed the failure of their cities to clean the streets out front. It is easy to imagine many more neglected tasks of this kind and hard to imagine "work" that can yield more satisfaction to the community and more fulfillment to many individuals. These examples underline further the lesson that Americans have been learning in recent years: effective political action can take many forms; self-government is not conducted by plebiscite.

The ethic that downgrades public employment and scorns the welfare recipient creates in its own image the social order labeled by Edward C. Banfield of Harvard University as the "lower class." In all humanity, first priority should be accorded to measures that will save the children from such dehumanization. That is not likely to be accomplished, however, by Banfield's own modest proposal that welfare mothers sell their babies to working-class or middle-class families. The revolutionary rhetoric of the rebellious young men and women of the slums is sometimes frightening to hear. But it is the most encouraging sound ever to come from those places, and it is perfectly intelligible when heard in the context of talk like Banfield's in the surrounding larger society.

The first demand on the agenda of these young people—so long as they have not completely despaired—is for justice. Why can't they have justice? This is one public employment, along with the military, that is accorded popular respect. Yet our country spends less than one-third of 1 percent of its public budget on the administration of justice. In the Manhattan felony court, 85 percent of the traffic is made up of blacks and other non-whites hauled blocks and miles from their homes to fester without limit of time in the city house of detention, awaiting trial under monstrous conditions to which they have drawn public attention only by rioting. The slum cannot evolve a true community life until justice to its residents can be administered within its boundaries and by juries of their "peers."

The American experience gives us no ground for placing a higher expectation of rationality in the political process than in the market. It is constitutionally evident, however, that the private transaction between buyer and seller cannot be trusted to activate and organize twentieth-century technology. The interests of too many other parties are too intimately engaged. Equity calls for a wider forum for an increasing number of market decisions. That forum we must learn to provide in our politics, especially state and local politics. Demonstrations against pollution and despoliation of the countryside come in belated recognition of the profound social and cultural consequences of the millions of private transactions that committed our transportation to the internal combustion engine and the highway.

The American landscape in the last third of the twentieth century demonstrates equally that the market process does not, of itself, automatically secure the benefits of technology. No basic research or radical invention is required to create public transportation systems that will make our cities habitable again. For the public benefit of clearing automobiles from the streets, public transportation could be offered free. Such radical substitution of the political for the market process probably still exceeds the grasp of the American democracy. On the other hand, the ecology movement has brought the political process more deeply into such market decisions as the siting of electrical power plants, the laying down of pipelines, and the harvesting of the national forest. None of these decisions can now be made—if the new ecologists stay on their toes—without reference to the interests of the wider public and future generations.

Nothing less than the resolution of the paradox of the aggregate—the accommodation of private convenience and satisfaction to the general welfare—can bring industrial civilization into happy adjustment to the resources of the earth and the cycles of the biosphere. A steady state of human population and its consumption of materials and energy, to paraphrase J. S. Mill, does not imply a stationary state of human improvement. The GNP can even go on growing; its greatest growth today reflects value of such nonmaterial and human inputs as sales

and services, communications, technology, teaching, and healing. Our planet, in harmony with the cosmos, is finite but
unbounded.

In the civilization of high technology, no man can improve
his lot at the expense of others; what diminishes them must
also, sooner rather than later, diminish him. The decisions that
bear most closely upon the well-being and expectations of the
individual are those that engage the well-being and expectations
of all. The goods and services, from libraries to landscapes, that
most enrich the lives of each and all are those that are not
diminished by the enjoyment of them. The material means are
at hand for the building of a society that will not know war or
waste or want.

GERARD PIEL, the publisher of *Scientific American*, was born at Woodmere, Long Island, in 1915 and was educated at Harvard College, where he received his A.B. *magna cum laude* in 1937. After serving as science editor of *Life* Magazine from 1939 until 1945, he became assistant to the president of the Henry Kaiser Company and associated companies, and worked in that capacity in 1945 and 1946. In 1948, in association with two colleagues, he launched the new *Scientific American,* recognizing at that time that communication among scientists and the general public would be one of the most pressing needs of postwar America. Mr. Piel, who holds three honorary degrees— Doctor of Science from Lawrence College (1956) and from Colby College (1960) and Doctor of Letters from Rutgers University (1961)—is also a Fellow of the American Academy of Arts and Sciences and of the American Association for the Advancement of Science, as well as a Trustee of the American Museum of Natural History. He was awarded the UNESCO Kalinga Prize in 1962, the George K. Polk Award in 1964, the Bradford Washburn Award in 1966, and the Arches of Science Award in 1969. Mr. Piel lives in New York, is married to a successful New York attorney, and has two children.

This book is set in Electra, a Linotype face designed by W. A. Dwiggins (1880–1956), who was responsible for so much that is good in contemporary book design. Although much of his early work was in advertising and he was the author of the standard volume *Layout in Advertising*, Mr. Dwiggins later devoted his prolific talents to book typography and type design and worked with great distinction in both fields. In addition to his designs for Electra, he created the Metro, Caledonia, and Eldorado series of type faces, as well as a number of experimental cuttings that have never been issued commercially.

Electra cannot be classified as either modern or old-style. Is is not based on any historical model, nor does it echo a particular period or style. It avoids the extreme contrast between thick and thin elements that marks most modern faces, and attempts to give a feeling of fluidity, power, and speed.

Composed, printed, and bound by H. Wolff, Inc., New York, New York.

Typography and binding design by CLINT ANGLIN.